T0214055

Studies in Systems, Decision and Control

Volume 390

Series Editor

Janusz Kacprzyk, Systems Research Institute, Polish Academy of Sciences, Warsaw, Poland

The series "Studies in Systems, Decision and Control" (SSDC) covers both new developments and advances, as well as the state of the art, in the various areas of broadly perceived systems, decision making and control–quickly, up to date and with a high quality. The intent is to cover the theory, applications, and perspectives on the state of the art and future developments relevant to systems, decision making, control, complex processes and related areas, as embedded in the fields of engineering, computer science, physics, economics, social and life sciences, as well as the paradigms and methodologies behind them. The series contains monographs, textbooks, lecture notes and edited volumes in systems, decision making and control spanning the areas of Cyber-Physical Systems, Autonomous Systems, Sensor Networks, Control Systems, Energy Systems, Automotive Systems, Biological Systems, Vehicular Networking and Connected Vehicles, Aerospace Systems, Automation, Manufacturing, Smart Grids, Nonlinear Systems, Power Systems, Robotics, Social Systems, Economic Systems and other. Of particular value to both the contributors and the readership are the short publication timeframe and the world-wide distribution and exposure which enable both a wide and rapid dissemination of research output.

Indexed by SCOPUS, DBLP, WTI Frankfurt eG, zbMATH, SCImago.

All books published in the series are submitted for consideration in Web of Science.

More information about this series at http://www.springer.com/series/13304

Muammer Catak · Tofigh Allahviranloo ·
Witold Pedrycz

Probability and Random Variables for Electrical Engineering

Probability: Measurement of Uncertainty

 Springer

Muammer Catak
Electrical Engineering, College
of Engineering and Technology
American University of the Middle East
Kuwait, Kuwait

Tofigh Allahviranloo
Department of Applied Mathematics
Bahcesehir University
Istanbul, Turkey

Witold Pedrycz
Department of Electrical and Computer
Engineering
University of Alberta
Edmonton, AB, Canada

ISSN 2198-4182   ISSN 2198-4190   (electronic)
Studies in Systems, Decision and Control
ISBN 978-3-030-82924-7   ISBN 978-3-030-82922-3   (eBook)
https://doi.org/10.1007/978-3-030-82922-3

This Springer imprint is published by the registered company Springer Nature Switzerland AG
The registered company address is: Gewerbestrasse 11, 6330 Cham, Switzerland

Dedicated to Our Families

Preface

Overview

This textbook entitled Probability and Random Variables for Electrical Engineering delivers a concise and carefully structured introduction to probability and random variables. It is aimed to build a linkage between the theoretical conceptual topics and the practical applications, especially in the undergraduate engineering area. The book motivates the student to gain a full understanding of the fundamentals of probability theory and help acquire working problem-solving skills and apply the theory to engineering applications. Each chapter includes solved examples at varying levels (both introductory and advanced) in addition to problems that demonstrate the relevance of the probability and random variables in engineering.

Chapter Descriptions

In Chap. 1, the Set Theory is discussed in detail. Thereafter, the probability theory is introduced using axiomatic terms based on the set theory. In Chap. 2, the general concept of the probability theory is introduced via random variables and the corresponding probability distribution functions and the probability density functions considering continuous-time random variables. Chapter 3 is dedicated to discuss discrete time random variables since, in many applications, a random variable should be defined in discrete form as a result of its nature being discrete in time. In addition, the mixed random variables, which are the combination of both continuous and discrete cases, are presented. In practical applications, a random variable may consist of both discrete and continuous probability density functions in order to give a better representation of a physical system. In Chap. 4, the multiple random variables and the joint probability functions are discussed. Moreover, the topics of marginal probability functions, statistical independency, and functions of random variables are explained. Statistical characteristics of random variables, such as expected value, mean, variance, standard deviation, moments, and moment generating functions are explored in Chap. 5.

In Chap. 6, we focused on the random processes. Correlation and covariance functions are introduced. Furthermore, the Gaussian and Poisson random processes are presented in detail. Spectral analysis of random processes is concerned in Chap. 7 for both continuous and discrete random processes. Chapter 8 is on the discussion of the linear systems with random inputs. System properties, frequency response, statistical analysis of random outputs, and noisy inputs are focused by means of brief discussions. Random samples, random matrices, and correlation and covariance matrices are covered in Chap. 9.

Kuwait, Kuwait   Muammer Catak
Istanbul, Turkey   Tofigh Allahviranloo
Alberta, Canada   Witold Pedrycz
June 2021

Contents

Part I
Concepts of Probability Theory

Chapter 1
Introduction

This chapter discusses the set theory in detail. The probability theory will be introduced in terms of set theory.

1.1 Set Theory

Although ancient Greek mathematicians established the fundamentals of the set theory, the origin of the set theory lies far beyond. If the mathematics is defined an abstract language to understand the real world, the set theory could be seen a tool to help classification of the world by means of mathematics.

Variable is a symbol that is certainly defined but has not a fixed value. For instance, x could be a variable assigned the number of students in a class, or could be the real numbers between 0 and 1. A set is defined as any range of variable. We may consider discrete variables such as number of students in a class; or continuous variables such as the angles (in radians) between 0 and 2π.

An *element*, which cannot be further fragmented, is the basic part of a set. Let x be the variable to denote the days of the week. Then the related set is written as; $\beta = \{Monday, Tuesday, Wednesday, Thursday, Friday, Saturday, Sunday\}$. The sets, which are made up any combination of the elements of a given set, are called *subsets*. For instance,

$$\beta_1 = \{Monday\}, \beta_2 = \{Tuesday, Wednesday\}$$

are subsets of β, and they are denoted as;

$$\beta_1 \subset \beta \quad \text{and} \quad \beta_2 \subset \beta$$

M. Catak et al., *Probability and Random Variables for Electrical Engineering*, Studies in Systems, Decision and Control 390, https://doi.org/10.1007/978-3-030-82922-3_1

The *empty set*, which has no elements, is called *null set*, and it is expressed as

$$\beta_N = \{\} = \varnothing$$

The *universal set*, Ω, contains all well-defined elements.

1.2   Set Operations

Let E_1 and E_2 are two subsets of the universal set, Ω.

Union   The union operation results in a new set containing all elements of E_1 and E_2. It is denoted as;

$$E_1 \cup E_2$$

Let
$$E_1 = \{1, 2, 3\} \quad \text{and} \quad E_2 = \{2, 4, 5\},$$

then
$$E_1 \cup E_2 = \{1, 2, 3, 4, 5\}$$

The union operation can be generalized more than two sets such as;

$$\cup_{i=1}^{n} E_i = E_1 \cup E_2 \cup \cdots \cup E_n$$

The possible union operations of two sets are illustrated in Fig. 1.1.

Intersection   The intersection operation, which is shown as

$$E_1 \cap E_2,$$

results in a new set consists of all elements that are both in E_1 and E_2. The generalized intersection operation for n-dimensional case can be shown as;

$$\cap_{i=1}^{n} E_i = E_1 \cap E_2 \cap ... \cap E_n$$

The possible intersection operations of two sets are illustrated in Fig. 1.2.

Complement   The complement of a set E, which is notated as E^C, contains all the elements of the universal set apart from the elements belongs to E.
A set E and its complement E^C are illustrated in Fig. 1.3.
One can easily deduce that

• The union of E and E^C equals to the universal set, Ω.

$$E \cup E^C = \Omega$$

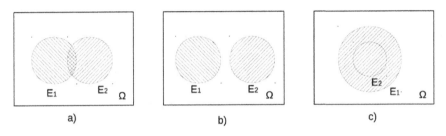

Fig. 1.1 Union operation of two sets

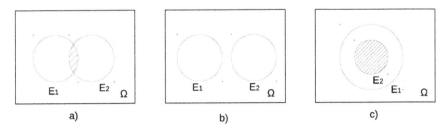

Fig. 1.2 Intersection operation of two sets

Fig. 1.3 Illustration of a set
E and its complement E^C

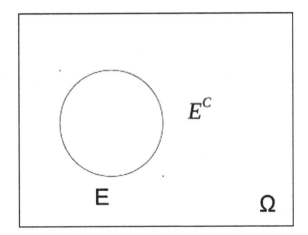

- The intersection of E and E^C is a null set, $\varnothing$.

$$E \cap E^C = \varnothing$$

Example 1.1 Sketch the following sets in $\Re^2$

(a) $x \geq 2$
(b) $y < 3$

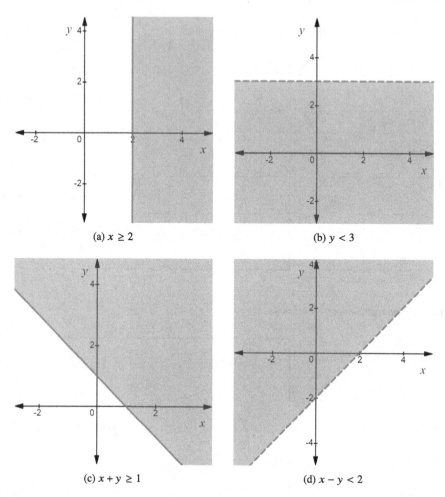

Fig. 1.4 Solution to Example 1.1

(c) $x + y \geq 1$
(d) $x - y < 2$.

Solution 1.1 The graphical illustrations of the sets are shown in Fig. 1.4. The continuous line means that the line is included by the set, while the dashed line means that the line is not included by the set.

1.3 Sample Space and Events

Consider an experiment whose outcomes cannot be predicted. For instance, if you flip a fair coin, nobody can certainly say that it comes "$Head(H)$" or "$Tail(T)$". The set includes all the possible outcomes of the experiment is called *sample space*. For instance;

- The sample space of the experiment of flipping a fair coin is;

$$S = \{H, T\}$$

- The sample space of the experiment of throwing a fair dice is;

$$S = \{1, 2, 3, 4, 5, 6\}$$

- The sample space of the experiment of flipping two fair coins is;

$$S = \{HH, HT, TH, TT\}$$

- The sample space of the experiment, where the sum of the outcomes of two thrown fair dices observed, is;

$$S = \{2, 3, 4, 5, 6, 7, 8, 9, 10, 11, 12\}$$

Any subset of the sample space is called *event*. For instance, if two fair dices are thrown $E = \{(1, 1), (1, 2), (1, 3), (1, 4), (1, 5), (1, 6), (2, 1), (3, 1), (4, 1), (5, 1), (6, 1)\}$ are the events that at least one of them comes out 1.

Example 1.2 Let an unfair dice such as two sides are marked 1, three sides are marked 2, and one side is marked 4 (i.e. $3, 5, 6$ are missing). Write down all the possible outcomes (sample space) of the experiment by throwing the dice once.

Solution 1.2 Since only the numbers of $\{1, 2, 4\}$ are marked on the sides of the unfair dice, the sample space is written as;

$$S = \{1, 2, 4\}$$

The sample space of the experiment of taking place by throwing the dice twice is listed as;

$$S = \{(1, 1), (1, 2), (1, 4), (2, 1), (2, 2), (2, 4), (4, 1), (4, 2), (4, 4)\}.$$

1.4 Probability Axioms

Let define a function $P(.)$ that maps variables from the sample space to the range of $[0, 1]$. Probability of an event E, which is the quantitative measurement of how likely E occurs as outcome of an experiment, is denoted as $P(E)$.

Axiom 1 Nonnegativity. *The probability of an event E must lie between 0 and 1.*

$$0 \leq P(E) \leq 1.$$

Axiom 2 Normalization. *The probability of the sample space equals 1, since it contains all the possible outcomes of an experiment.*

$$P(S) = 1.$$

Axiom 3 Mutually Exclusive. *If the intersection of two events, let say E_1 and E_2, is null set, then E_1 and E_2 are called mutually exclusive events. If E_is are mutually exclusive events for $i = 1, 2, \ldots, n$; the probability of the union of E_is equals to the summation of the probabilities of E_is.*

$$P(\cup_{i=1}^{n} E_i) = \sum_{i=1}^{n} P(E_i).$$

1.5 Probability of the Equally Likely Outcomes

The probability of equally likely outcomes of an experiment is calculated as the division of the number of elements of the desired event by the number of elements of the sample space.

$$P(E) = \frac{\text{Number of elements of } E}{\text{Number of elements of } S}$$

For instance, considering throwing a fair dice

$$P(1) = P(2) = P(3) = P(4) = P(5) = P(6) = \frac{1}{6}.$$

Example 1.3 Let an urn consists of 4 white, 3 black, and 5 red balls.

(a) What is the probability of that the randomly selected ball is white?
(b) What is the probability of that the randomly selected 3 balls are white, black, and red, respectively?
(c) What is the probability of that the randomly selected 3 balls are white, black, and red without ordering?

Solution 1.3 (a) There are 4 white balls, and there are 12 balls in total. Hence, the probability of that the randomly selected ball is white can be calculated as;

$$P(W) = \frac{4}{12} = \frac{1}{3}$$

(b) This case is likely to the conditional probability,

$$P(R|B|W) = \frac{4}{12}\frac{3}{11}\frac{5}{10} = \frac{1}{22}$$

(c) The total number of the outcomes is the number of selection ways of 3 balls within 12 balls, such as

$$\binom{12}{3} = \frac{12!}{3!(12-3)!}$$

The desired event is the combination the selecting 1 ball within 4 white balls, 1 ball within 3 black balls, and 1 ball within 5 red balls.

$$P(WBR) = \frac{\binom{4}{1}\binom{3}{1}\binom{5}{1}}{\binom{12}{3}} = \frac{3}{11}$$

1.6 Conditional Probability and Bayes' Theory

The conditional probability of the event B under the condition that the event A has already taken place. It is denoted as $P(B|A)$,and it equals to the probability of the intersection of the event A and the event B divided by the probability of the event A. Such as;

$$P(B|A) = \frac{P(B \cap A)}{P(A)}$$

where $P(A) \neq 0$.

The events are called exhaustive if the union of them equals to the sample space. The difference between the complements and the exhaustive events might be explained in the following way: For the complement case, the union of the event E and its complement E^C equals to the sample space; but for the exhaustive case, the number of the events, whose union operation equals to the sample space might be more than two.

Let $A_i, i = 1, 2, 3, \ldots, n$ be a set of mutually exclusive (i.e. disjoint), and exhaustive events defined on the sample space S.

(i) $A_i \cap A_j = \emptyset$ (mutually exclusive)
(ii) $\cup_{i=1}^{n} A_i = \Omega$ (exhaustive)

The Bayes theory, which is an application of the conditional probability, can be expressed as;

$$P(A_j|B) = \frac{P(B|A_j)P(A_j)}{\sum_{i=1}^{n} P(B|A_i)P(A_i)}$$

The numerator of the equation equals $P(B \cap A_j)$, and the denominator of the equation is $P(\sum P()B \cap A_i))$. That means the probability of being the event A_j under the condition that the event B has already occurred.

Example 1.4 Let an information is sent by means of a hundred data packets. The probability of being defected data packet is given the probability of 0.01. An information is accepted as correctly sent if at most 3 data packets are defected. What is the probability that an information is correctly sent?

Solution 1.4 Let $P(A)$ is denoted the probability of a data packet being defected, and $P(A^C)$ is the probability of a data packet being non-defected. $P(A)$ is given as 0.01, hence $P(A^C)$ must be 0.99. The desired event consists of the following outcomes,

	Outcome 1	Outcome 2	Outcome 3	Outcome 4
# of defected	0	1	2	3
# of non-defected	100	99	98	97

Therefore, the probability that an information is correctly sent is;

$$(0.99)^{100} \times (0.01)^0 + (0.99)^{99} \times (0.01)^1 + (0.99)^{98} \times (0.01)^2$$
$$+ (0.99)^{97} \times (0.01)^3 \approx 0.3698.$$

Example 1.5 What is the probability that at least one head comes up after flipping a fair coin four times?

Solution 1.5 It is known that the probabilities of coming head (H) and coming tail (T) are equally likely and;

$$P(H) = P(T) = \frac{1}{2}$$

The desired event is the *complement* of the event such that no head comes up. Therefore

$$P(\text{at least one head comes up}) = 1 - P(\text{no head comes up})$$

which means four tails come up with the probability of $(\frac{1}{2})^4 = \frac{1}{16}$. Hence;

$$P(\text{at least one head comes up}) = 1 - \frac{1}{16} = \frac{15}{16}$$

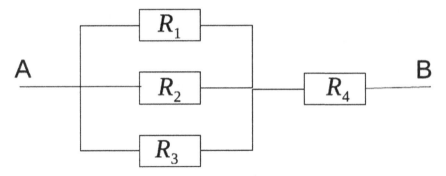

Fig. 1.5 A Simple resistive circuit for Example 1.6

Example 1.6 A basic circuit composed of resistors is shown in Fig. 1.5. The probabilities of R_1, R_2, R_3, R_4 being broken are given as $0.1, 0.1, 0.1, 0.01$, respectively. Calculate the probability that the current flowing through the circuit is nonzero, i.e. there should be at least one way connecting the point A and the point B.

Solution 1.6 Let $P(R_i)$ be the probability corresponding to the event of R_i is broken. If R_1, R_2, R_4 and /or R_4 are broken, then no current passing through the circuit. This is the complement of the situation for which we try to find its probability. Hence;

$$P(R_1 \cap R_2 \cap R_3) \cup P(R_4) = P(R_1 \cap R_2 \cap R_3) + P(R_4) - P(R_1 \cap R_2 \cap R_3) \cap P(R_4)$$

which equals to
$$(0.1)^3 + 0.01 - (0.1)^3 \times (0.01) \approx 0.011$$

Therefore, the probability of a current passing through the circuit is $1 - 0.011 = 0.989$.

Example 1.7 There are three urns containing various colors of balls. Let the ratios of the red balls to the all balls in $urn - 1$, $urn - 2$, and $urn - 3$ be $\frac{1}{2}, \frac{2}{3}, \frac{3}{4}$, respectively. A fair die is thrown, then an urn is selected regarding to the outcome of the die such as;

Outcome of the die	Selected urn
1,2,3	1
4	2
5,6	3

Solution 1.7 According to the outcome of the die, the probability of selecting $urn - 1$ is

$$\frac{3}{6} = \frac{1}{2}$$

To select randomly a red ball from $urn - 1$ is given as $\frac{1}{2}$. Hence, the probability of selecting a red ball from $urn - 1$ at the end of the experiment can be calculated as;

$$\frac{1}{2}\frac{1}{2} = \frac{1}{4}$$

Similarly, the probability of to select a red ball from $urn - 2$ is

$$\frac{1}{6}\frac{2}{3} = \frac{1}{9}$$

and the probability of to select a red ball from $urn - 3$ is

$$\frac{2}{6}\frac{3}{4} = \frac{1}{4}$$

since, it is known that a red ball has been already drawn, the probability that the red ball is drawn from the $urn - 1$, $urn - 2$, and $urn - 3$ can be calculated as;

$$\text{for } urn - 1 \quad \frac{\frac{1}{4}}{\frac{1}{4} + \frac{1}{9} + \frac{1}{4}} = \frac{9}{22}$$

$$\text{for } urn - 2 \quad \frac{\frac{1}{9}}{\frac{1}{4} + \frac{1}{9} + \frac{1}{4}} = \frac{4}{22}$$

$$\text{for } urn - 3 \quad \frac{\frac{1}{4}}{\frac{1}{4} + \frac{1}{9} + \frac{1}{4}} = \frac{9}{22}$$

1.7 Problems

1.1 Let three sets are defined as;

$$S_1 = \{(x, y) : x \leq y\}$$

$$S_2 = \{(x, y) : x - y \leq 3\}$$

$$S_3 = \{(x, y) : x + y > 5\}$$

Show the following sets on the xy-plane

(a) $S_1 \cup S_2$
(b) $S_1 \cup S_3$
(c) $S_2 \cap S_3$
(d) $S_1 \cap S_2 \cap S_3$.

1.2 Consider the following given sets

$$S_1 = \{(x, y, z) : x \geq y \geq z\}$$

$$S_2 = \{(x, y, z) : x \geq y + 2, z = 3\}$$

Explain the following sets in the $\mathfrak{R}^3$

(a) $S_1 \cup S_2$
(b) $S_1 \cap S_3$.

1.3 Two fair coins and one unfair coin with "Head" of both sides are thrown. Write down all possible outcomes.

1.4 In a certain city, the last four digits of the car plates must be integers. What is the probability that randomly selected two cars;

(a) have the same last digits
(b) have different integer of last two digits
(c) both have even last digits.

1.5 Four balls are drawn from an urn including 4 black, 6 red, 5 yellow, 10 white colored balls without replacement. Calculate the probabilities that;

(a) at least one ball is black
(b) all of the four balls are yellow or black
(c) all of the four balls have different colors
(d) repeat the parts (a)–(c) under the case of replacement.

1.6 Let the possible grades of the probability course be A, B, C, and F (fail) with the probabilities of $0.1, 0.3, 0.4$, and 0.2, respectively. If there are 30 students attended to the course, find the following probabilities;

(a) at least one student gets A
(b) exactly ten students get B
(c) a randomly selected student be failed.

1.7 Three fair die are thrown. Calculate the probabilities that;

(a) sum of upcoming numbers is 18
(b) at least one of the upcoming numbers is 2.

1.8 The probability of being defected of a capacitor is 0.01. What is the probability that at least one defected capacitor in a 100 of group.

1.9 The possible transportation ways of a person, who lives in the city A and works in the city C are shown in Fig. 1.6.

(a) Write down all the possible combinations to transport from A to C.

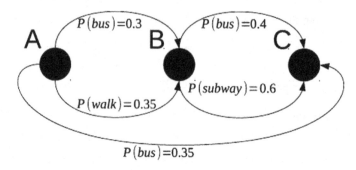

Fig. 1.6 Probability values used in the Problem 1.9

(b) Calculate the probability that the person takes subway at least once in a day (remind the return).

1.10 A grandfather and his grandson go to fishing. The grandfather catches the fish number of 1–5 with the probabilities of 0.1, 0.2, 0.2, 0.3, 0.2, respectively. The grandson catches the fish number of 2–6 with the probabilities of 0.2, 0.2, 0.2, 0.3, 0.1, respectively. Calculate the probability that they return to their home with more than 4 fishes.

1.11 Let a TV unit have 3 main integrated circuits (ICs), namely A, B, and C. the probabilities of being defected of ICs are given as;

$$P(A : \text{defected}) = 0.1$$
$$P(B : \text{defected}) = 0.15$$
$$P(C : \text{defected}) = 0.18$$

The refund-or-change policy of the company for a brand new TV unit is like that: (i) If only one ICs is defected, the company changes it with a brand new one. (ii) If two ICs are defected, the customer may change it or the company may refund it with equally likely probabilities. (iii) If three ICs are defected the company definitely refund it.

(a) What is the probability that a brand new TV unit has at least one defected ICs?
(b) If the company refunded a product, what is the probability that the TV unit has only two defected ICs?

1.12 Let imagine a dice with infinitely many sides that are labeled by countable positive integers such as $1, 2, 3, \ldots$

(a) What is the probability that the outcome of the thrown dice is 1?
(b) What is the probability that the outcome is an odd number?

1.13 Let imagine two die with infinitely many sides. The sides of the first dice are labeled by positive odd integers such as $1, 3, 5, \ldots$, and the sides of the second dice are labeled by the positive even integers such as $2, 4, 6, \ldots$ If these two die are thrown;

(a) What is the probability that the sum of the outcomes is an even number?
(b) If the outcome of the first dice is odd, calculate the probability that the sum of the outcomes is even?

Chapter 2
Continuous Random Variables

The basic theory of the probability has been discussed in Chap. 1. That knowledge based on the set theory is essential to understand the notion of probability. However, it lacks applicability in many cases. For instance, let consider that very large number of events is under investigation. Calculations of probabilities of any arbitrary outcomes would require a great deal of effort. For this reason, the general concept of the probability theory is developed by means of random variables and corresponding probability functions, namely the probability distribution function and the probability density function. In general, a function maps elements of a well-defined set to another well-defined set. Let $x \in A$ and $y \in B$, a function $f(x)$ defines the transformation from set A to set B, and it is denoted as;

$$f : x \rightarrow y$$

A random variable, X, is a function related with the outcomes of a certain random experiment. X maps all the possible outcomes belonging the sample space S into the set of the real numbers $\Re$. Let a and b be any real numbers, the probabilities of X according to its range are listed in Table 2.1.

2.1 The Probability Distribution Function (PDF)

The probability distribution function (PDF) states that the probability of the random variable X is less than or equal to a real number such as a.

$$F_X(a) = P(X \leq a) \text{ for } -\infty < a < \infty$$

© The Author(s), under exclusive license to Springer Nature Switzerland AG 2022
M. Catak et al., *Probability and Random Variables for Electrical Engineering*,
Studies in Systems, Decision and Control 390,
https://doi.org/10.1007/978-3-030-82922-3_2

where $F_X(x)$ is the probability distribution function of the random variable X. Since, it is a cumulative function, it is also called the cumulative distribution function (CDF).

Properties of $F_{X(x)}$

Based on the axioms of the probability theory introduced in Chap. 1, the properties of $F_X(x)$ are enumerated as;

1. $0 \le F_X(x) \le 1$ for all x values.
2. $F_X(x)$ is a non-decreasing function, since it is a cumulative function,

$$F_X(a) \le F_X(b) \quad \text{if} \quad a < b$$

3.

$$\lim_{x \to \infty} F_X(x) = \lim_{a \to \infty} P(X \le a) = 1$$

4.

$$\lim_{x \to -\infty} F_X(x) = \lim_{a \to -\infty} P(X \le a) = 0$$

5. $P(X = a) = 0$, where a is certain constant. This property is valid only for the continuous random variables, that means $F_X(x)$ is a continuous function on $\Re$. It may not be valid for the discrete and mixed random variables (those will be discussed in the next chapter).
6. Using property 5;

$$P(a < X \le b) = P(a < X < b) = P(a \le X \le b) = P(a \le X < b)$$
$$= F_X(b) - F_X(a)$$

7.

$$P(X > a) = P(x \ge a) = 1 - F_X(a)$$

2.2   The Probability Density Function (PDF)

The change of the probability distribution function $F_X(x)$ with respect to the variable x is called the corresponding probability density function, $f_X(x)$;

$$f_X(x) = \frac{d F_X(x)}{dx}$$

Table 2.1 Events and the corresponding probability notations

Event	Corresponding probability notation
(i) $x = a$	$P(X = a)$
(ii) $a < x \leq b$	$P(a < X \leq b)$
(iii) $x \leq b$	$P(X \leq b)$
(iv) $x \geq b$	$P(X \geq b)$

Properties of $f_{X(x)}$

1. Since $F_X(x)$ is a non-decreasing function, then $f_X(x)$ must be greater than or equal to 0.

$$f_X(x) \geq 0$$

2. The area under the curve of $f_X(x)$ equals to 1 in analogy with $P(S) = 1$.

$$\int_{-\infty}^{\infty} f_X(x)dx = 1$$

3. Considering for any given $f_X(x)$, $F_X(x)$ can be calculated as;

$$F_X(x) = \int_{-\infty}^{x} f_X(\tau)d\tau$$

where τ is a dummy variable.

4.

$$P(a < X \leq b) = F_X(b) - F_X(a) = \int_{a}^{b} f_X(x)dx$$

Example 2.1 Let a random variable X has the following probability density function;

$$f_X(x) = \begin{cases} ax(1 - x^2) & 1 \leq x \leq 2 \\ 0 & \text{elsewhere} \end{cases}$$

(a) Find a such that $f_X(x)$ is a proper probability density function.
(b) Show a plot $F_X(x)$
(c) Calculate the probabilities of $P(0 < X \leq 5)$, $P(0.25 < X < 1)$, and $P(X = 1.5)$

Solution 2.1 (a) According to the property 2 described above;

$$\int_{-\infty}^{\infty} f_X(x)dx = 1$$

$$\int_{1}^{2} ax(1-x^2)dx = 1$$

$$(\frac{a}{2}x^2 - \frac{a}{4}x^4) \,|_1^2 = 1$$

$$(2a - 4a) - (\frac{a}{2} - \frac{a}{4}) = 1$$

$$\text{then} \quad a = -\frac{4}{9}$$

Please note that, a is negative parameter, but $f_X(x)$ is positive as it must be.
(b) Firstly, the real line is divided into subintervals based on the domain of $f_X(x)$
illustrated in Figs. 2.1 and 2.2.

- Case I $x < 1$

$$F_X(x) = \int_{-\infty}^{x} f_X(\tau)d\tau = 0$$

- Case II $1 \le x \le 2$

$$F_X(x) = \int_{-\infty}^{x} f_X(\tau)d\tau$$

$$F_X(x) = \int_{-\infty}^{x} -\frac{4}{9}\tau(1-\tau^2))d\tau$$

$$F_X(x) = (-\frac{2}{9}\tau + \frac{1}{9}\tau^4) \,|_1^x$$

$$F_X(x) = \frac{1}{9}x^4 - \frac{2}{9}x^2 + \frac{1}{9}$$

- Case III $x > 2$

$$F_X(x) = \int_{-\infty}^{x} f_X(\tau)d\tau = \int_{1}^{2} f_X(\tau)d\tau = 1$$

Thereby;

$$F_X(x) = \begin{cases} 0 & x < 1 \\ \frac{1}{9}x^4 - \frac{2}{9}x^2 + \frac{1}{9} & 1 \le x \le 2 \\ 1 & x > 2 \end{cases}$$

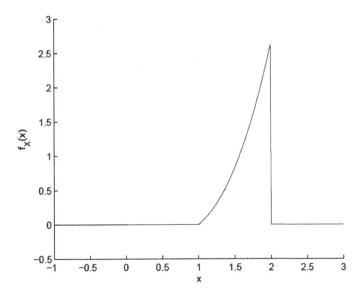

Fig. 2.1 Probability density function (Example 2.1)

Fig. 2.2 The domain of $f_X(x)$(Example 2.1)

(c)

$$P(0 < X \le 5) = F_X(5) - F_X(0) = 1$$

$$P(0.25 < X < 1) = F_X(1) - F_X(0.25) = 0$$

$$P(X = 1.5) = \lim_{\epsilon \to 0} F_X(1.5 + \epsilon) - F_X(1.5 - \epsilon) = 0$$

Example 2.2 Let the probability density function of a random variable X be described as;

$$f_X(x) = \begin{cases} Asin(\frac{\pi x}{8}) & 0 \le x \le 8 \\ 0 & \text{elsewhere} \end{cases}$$

(a) Find A such that $f_X(x)$ is a proper probability density function.
(b) Find and plot $F_X(x)$
(c) Calculate the probabilities of $P(0 < X \le \pi)$, $P(X \ge 2\pi)$, $P(X = \frac{\pi}{2})$

Solution 2.2 (a) According to the property 2 described above;

$$\int_{-\infty}^{\infty} f_X(x)dx = 1$$

$$\int_{0}^{8} A\sin(\frac{\pi x}{8})dx = 1$$

$$\frac{-8A}{\pi}\cos(\frac{\pi x}{8}) \mid_{0}^{8} = 1$$

$$\frac{-8A}{\pi}(\cos(\pi) - \cos(0)) = 1$$

$$\text{then} \quad A = \frac{\pi}{16}$$

(b) $F_X(x)$ can be easily calculated based on the domain of $f_X(x)$, such that (Figs. 2.3 and 2.4),

- Case I $x < 0$

$$F_X(x) = \int_{-\infty}^{x} f_X(\tau)d\tau = 0$$

- Case II $0 \le x \le 8$

$$F_X(x) = \int_{-\infty}^{x} f_X(\tau)d\tau$$

$$F_X(x) = \int_{1}^{x} \frac{\pi}{16}\sin(\frac{\pi \tau}{8}))d\tau$$

$$F_X(x) = (\frac{\pi}{16}(-\frac{8}{\pi}\cos(\frac{\pi \tau}{8}))) \mid_{0}^{x}$$

$$F_X(x) = \frac{1}{2} - \frac{1}{2}\cos(\frac{\pi x}{8})$$

- Case III $x > 8$

$$F_X(x) = \int_{-\infty}^{x} f_X(\tau)d\tau = \int_{1}^{8} f_X(\tau)d\tau = 1$$

Hence;

$$F_X(x) = \begin{cases} 0 & x < 0 \\ \frac{1}{2} - \frac{1}{2}\cos(\frac{\pi x}{8}) & 0 \le x \le 8 \\ 1 & x > 8 \end{cases}$$

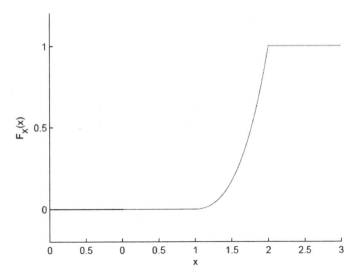

Fig. 2.3 Probability distribution function (Example 2.1)

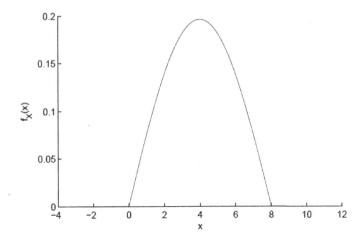

Fig. 2.4 Probability density function (Example 2.2)

(c)

$$P(0 < X \leq \pi) = F_X(\pi) - F_X(0) = 0.335$$

$$P(X \geq 2\pi) = 1 - F_X(2\pi) = 1 - 0.891 = 0.109$$

$$P(X = \frac{\pi}{2}) = \lim_{\epsilon \to 0} F_X(\frac{\pi}{2} + \epsilon) - F_X(\frac{\pi}{2} - \epsilon) = 0$$

2.3   Expected Value and Variance

Expected value, $E[X]$, and the variance, $Var[X]$, are two significant parameters to understand the behavior of a distribution. The expected value and the mean, μ_X, are widely used as synonym. But there is a slight difference between them, such that while the mean is referred for the average value of both a distribution and a randomly generated data relying on a probability distribution, the expected value is just used for a distribution. The expected value of a probability distribution is calculated as;

$$E[X] = \int_{-\infty}^{\infty} x f_X(x) dx$$

$Var[X]$ is a positive parameter that supplies information on how likely the related random variable variate around the mean. It is defined as;

$$Var[X] = \int_{-\infty}^{\infty} (x - E[X])^2 f_X(x) dx$$

The $(.)^2$ function is necessary in order to avoid cancellation of the positive and the negative variations. Then (Fig. 2.5),

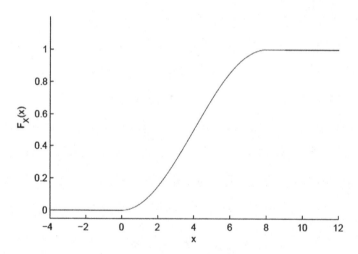

Fig. 2.5   Probability distribution function (Example 2.2)

$$Var[X] = \int_{-\infty}^{\infty} (x^2 - 2x E[X] + E[X]^2) f_X(x) dx$$

$$= \int_{-\infty}^{\infty} x^2 f_X(x) dx - 2E[X] \int_{-\infty}^{\infty} x f_X(x) dx + E[X]^2 \int_{-\infty}^{\infty} f_X(x) dx$$

since,

$$\int_{-\infty}^{\infty} x f_X(x) dx = E[X]$$

$$\int_{-\infty}^{\infty} f_X(x) dx = 1$$

then, it can be written as;

$$Var[X] = \int_{-\infty}^{\infty} x^2 f_X(x) dx - E[X]^2$$

The standard deviation, σ_X, is the square root of the variance. σ_X is commonly used to characterize the probability density functions instead of the variance,

$$\sigma_X = \sqrt{Var[X]}.$$

2.4 Common Continuous Probability Distribution Functions

In this section, some commonly used continuous probability distribution functions representing continuous random variables are introduced.

2.4.1 Uniform Distribution

If a random variable is equally likely probable in a finite interval, then it is called uniformly distributed. The probability density function has a constant value in that certain interval, and it is zero in the rest of the real line. According to the probability axioms;

$$\int_{-\infty}^{\infty} f_X(x) dx = \int_{a}^{b} c \, dx = 1$$

Hence;

$$c(b - a) = 1$$

$$c = \frac{1}{b - a}$$

The expected value of a uniform distribution is calculated as;

$$E[X] = \int_a^b cx\,dx = \frac{c}{2}x^2 \Big|_a^b = \frac{a+b}{2}$$

The variance of a uniform distribution is;

$$Var[X] = \int_a^b cx^2\,dx - E[X]^2$$

$$= \frac{c}{3}x^3 \Big|_a^b - \left(\frac{a+b}{2}\right)^2$$

$$= \frac{1}{12}(b-a)^2$$

The uniform distribution is generally used to represent a random variable for the cases such that;
– If there is no knowledge about the randomness behavior of the variable, that means, there is no rational justification in order to favour some possibilities.
– The random nature of the process might be illustrated by means of a uniform distribution within an acceptable error band.

Example 2.3 The waiting time at a bus stop is uniformly distributed between 0 and 5 min, that is at least one bus comes within a 5 min.

(a) Define and sketch the probability density function $f_X(x)$
(b) Find and plot the probability distribution function $F_X(x)$
(c) Calculate the following probabilities of $P(X \le 2)$, $P(1 < X \ge 2)$, $P(X = 2)$, $P(X > 3)$ (Fig. 2.6).

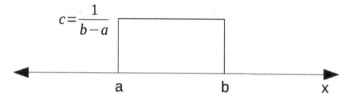

Fig. 2.6 An illustrative uniform probability density function

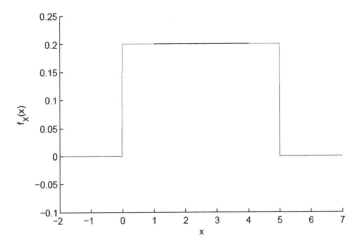

Fig. 2.7 Probability distribution function (Example 2.3)

Solution 2.3 (a) The probability density function is described as;

$$f_X(x) = \begin{cases} c & x < 0 \le 5 \\ 0 & \text{elsewhere} \end{cases}$$

Since;

$$\int_{-\infty}^{\infty} f_X(x)dx = \int_0^5 c\,dx = 1$$

Hence;

$$c(5 - 0) = 1$$

$$c = \frac{1}{5}$$

(b) $F_X(x)$ can be easily calculated based on the domain of $f_X(x)$, such that (Fig. 2.7)

- Case I $x < 0$

$$F_X(x) = \int_{-\infty}^{x} f_X(\tau)d\tau = 0$$

- Case II $0 \le x \le 5$

$$F_X(x) = \int_{-\infty}^{x} f_X(\tau)d\tau$$

$$F_X(x) = \int_{0}^{x} \frac{1}{5}d\tau$$

$$F_X(x) = \left(\frac{\tau}{5}\right) \Big|_0^x$$

$$F_X(x) = \frac{x}{5}$$

- Case III $x > 5$

$$F_X(x) = \int_{-\infty}^{x} f_X(\tau)d\tau$$

$$F_X(x) = \int_{0}^{5} \frac{1}{5}d\tau$$

$$F_X(x) = \left(\frac{\tau}{5}\right) \Big|_0^5$$

$$F_X(x) = 1$$

Hence;

$$F_X(x) = \begin{cases} 0 & x < 0 \\ \frac{x}{5} & 0 \leq x \leq 5 \\ 1 & x > 5 \end{cases}$$

(c)

$$P(X \leq 2) = F_X(2) = 0.4$$

$$P(1 < X \geq 2) = F_X(2) - F_X(1) = 0.4 - 0.2 = 0.2$$

$$P(X = 2) = \lim_{\epsilon \to 0} F_X(2 + \epsilon) - F_X(2 - \epsilon) = 0$$

$$P(X > 3) = 1 - F_X(3)1 - 0.6 = 0.4.$$

Example 2.4 A complex number can be represented in polar form such as $z = a + jb = re^{j\theta}$, where r is the magnitude and θ is the phase angle. By the definition, $0 \leq r \leq \infty$ and $0 \leq \theta \leq 2\pi$. What is the probability that a randomly assigned complex number has the phase angle between $\frac{\pi}{4}$ and $\frac{\pi}{3}$?

Solution 2.4 The probability density function of the phase angle in Example 2.4 is given in Fig. 2.9.

$$P\left(\frac{\pi}{4} \leq X \leq \frac{\pi}{3}\right) = \int_{\frac{\pi}{4}}^{\frac{\pi}{3}} \frac{1}{2\pi}d\theta = \frac{\theta}{2\pi} \Big|_{\frac{\pi}{4}}^{\frac{\pi}{3}} = \frac{1}{24}$$

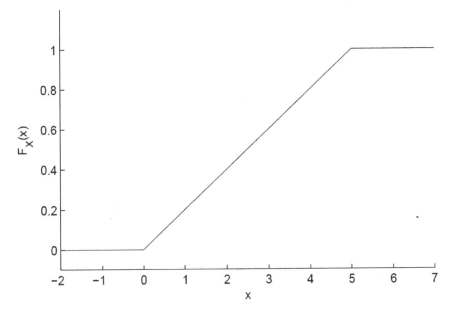

Fig. 2.8 Probability distribution function (Example 2.3)

Therefore, the probability that a randomly assigned complex number having the phase angle between $\frac{\pi}{4}$ and $\frac{\pi}{3}$ is almost $0.042 = 4.2\%$.

2.4.2 Normal (Gaussian) Distribution

A normal probability density function is widely used to represent a random variable in all studies such as engineering, social sciences, finance. Two main parameters, namely the mean (μ_X), and the standard deviation (σ_X), characterize a normal distribution. A univariate normal probability density function is written as;

$$f_X(x) = \frac{1}{\sqrt{2\pi\sigma_X^2}} e^{\frac{-1}{2}\left(\frac{x-\mu_X}{\sigma_X}\right)^2}$$

A normal distribution is usually denoted as $N(\mu_X, \sigma_X)$.
The probability distribution function is calculated as (Fig. 2.8);

$$F_X(x) = \frac{1}{\sqrt{2\pi\sigma_X^2}} \int_{-\infty}^{x} e^{\frac{-1}{2}\left(\frac{\tau-\mu_X}{\sigma_X}\right)^2} d\tau$$

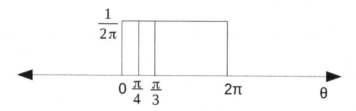

Fig. 2.9 Uniform distribution of the phase angle (Example 2.4)

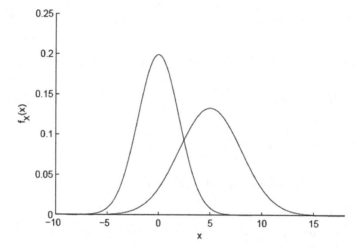

Fig. 2.10 Normal distribution with ($\mu_X = 0$, $\sigma_X = 2$) and ($\mu_X = 5$, $\sigma_X = 3$)

Let $\xi = \frac{\tau - \mu_X}{\sigma_X}$, then $d\xi = \frac{1}{\sigma_X} d\tau$. Hence,

$$F_X(x) = \frac{1}{\sqrt{2\pi}} \int_{-\infty}^{\frac{x - \mu_X}{\sigma_X}} e^{\frac{-1}{2}\xi^2} d\xi$$

The integral part of the equation cannot be solved analytically. Either numerical techniques, or series approximations are employed to compute this the integral.

The error function, erf (x) or commonly $\Phi(x)$ in engineering, is described as (Figs. 2.9 and 2.10);

$$\Phi(x) = \frac{1}{\sqrt{2\pi}} \int_{-\infty}^{x} e^{\frac{-1}{2}\xi^2} d\xi$$

Therefore, the value of the probability distribution function at a certain point is calculated by means of the error function. Such as;

$$F_X(x) = \Phi\left(\frac{x - \mu_X}{\sigma_X}\right)$$

The values of $\Phi(x)$ are given in Table 2.2 for $x \geq 0$. For the negative x values the following relationship is applied;

$$\Phi(-x) = 1 - \Phi(x).$$

Example 2.5 The amplitude of a noisy part of a signal is normally distributed with zero (Volt) mean and a variance of 4 (Volt^2). If the absolute value of the amplitude of the noise is > 1 Volt, then the signal is accepted as deterred. What is the probability that the signal is not deterred?

Solution 2.5

$$P(|X| < 1) = P(-1 < X < 1) = P\left(\frac{-1-\mu_X}{\sigma_X} < Z < \frac{1-\mu_X}{\sigma_X}\right)$$
$$P\left(\frac{-1}{2} < Z < \frac{1}{2}\right)$$

Referring to Table 2.2, the corresponding probability can be evaluated as;

$$P\left(\frac{-1}{2} < Z < \frac{1}{2}\right) = \Phi\left(\frac{1}{2}\right) - \Phi\left(\frac{-1}{2}\right) = \Phi\left(\frac{1}{2}\right) - \left(1 - \Phi\left(\frac{1}{2}\right)\right) = 0.383.$$

2.4.3 Exponential Distribution

The probability density function of the exponential distribution can be written as

$$f_X(x) = \begin{cases} \lambda e^{\lambda x} & \text{for } x \geq 0 \\ 0 & \text{elsewhere} \end{cases}$$

where λ is a proper positive constant. Hence, the probability distribution function is calculated in the following form;

$$F_X(x) = \begin{cases} 1 - e^{\lambda x} & \text{for } x \geq 0 \\ 0 & \text{elsewhere} \end{cases}$$

One can easily derive the expected value of the exponential distribution and the corresponding variance as;

$$E[X] = \frac{1}{\lambda}$$

$$\text{Var}[X] = \frac{1}{\lambda^2}$$

Exponential distribution is widely employed to model life time and waiting time processes.

Table 2.2 Values of $\Phi(x)$ for $0 \leq x \leq 3.29$ with 0.01 increments

x	0	0.01	0.02	0.03	0.04	0.05	0.06	0.07	0.08	0.09
0	0.5000	0.5040	0.5080	0.5120	0.5160	0.5199	0.5239	0.5279	0.5319	0.5359
0.1	0.5398	0.5438	0.5478	0.5517	0.5557	0.5596	0.5636	0.5675	0.5714	0.5754
0.2	0.5793	0.5832	0.5871	0.5910	0.5948	0.5987	0.6026	0.6064	0.6103	0.6141
0.3	0.6179	0.6217	0.6255	0.6293	0.6331	0.6368	0.6406	0.6443	0.6480	0.6517
0.4	0.6554	0.6591	0.6628	0.6664	0.6700	0.6736	0.6772	0.6808	0.6844	0.6879
0.5	0.6915	0.6950	0.6985	0.7019	0.7054	0.7088	0.7123	0.7157	0.7190	0.7224
0.6	0.7258	0.7291	0.7324	0.7357	0.7389	0.7422	0.7454	0.7486	0.7518	0.7549
0.7	0.7580	0.7612	0.7642	0.7673	0.7704	0.7734	0.7764	0.7794	0.7823	0.7852
0.8	0.7881	0.7910	0.7939	0.7967	0.7996	0.8023	0.8051	0.8079	0.8106	0.8133
0.9	0.8159	0.8186	0.8212	0.8238	0.8264	0.8289	0.8315	0.8340	0.8365	0.8389
1	0.8413	0.8438	0.8461	0.8485	0.8508	0.8531	0.8554	0.8577	0.8599	0.8621
1.1	0.8643	0.8665	0.8686	0.8708	0.8729	0.8749	0.8770	0.8790	0.8810	0.8830
1.2	0.8849	0.8869	0.8888	0.8907	0.8925	0.8944	0.8962	0.8980	0.8997	0.9015
1.3	0.9032	0.9049	0.9066	0.9082	0.9099	0.9115	0.9131	0.9147	0.9162	0.9177
1.4	0.9192	0.9207	0.9222	0.9236	0.9251	0.9265	0.9279	0.9292	0.9306	0.9319
1.5	0.9332	0.9345	0.9357	0.9370	0.9382	0.9394	0.9406	0.9418	0.9430	0.9441
1.6	0.9452	0.9463	0.9474	0.9485	0.9495	0.9505	0.9515	0.9525	0.9535	0.9545
1.7	0.9554	0.9564	0.9573	0.9582	0.9591	0.9599	0.9608	0.9616	0.9625	0.9633
1.8	0.9641	0.9649	0.9656	0.9664	0.9671	0.9678	0.9686	0.9693	0.9700	0.9706
1.9	0.9713	0.9719	0.9726	0.9732	0.9738	0.9744	0.9750	0.9756	0.9762	0.9767
2	0.9773	0.9778	0.9783	0.9788	0.9793	0.9798	0.9803	0.9808	0.9812	0.9817
2.1	0.9821	0.9826	0.9830	0.9834	0.9838	0.9842	0.9846	0.9850	0.9854	0.9857
2.2	0.9861	0.9865	0.9868	0.9871	0.9875	0.9878	0.9881	0.9884	0.9887	0.9890
2.3	0.9893	0.9896	0.9898	0.9901	0.9904	0.9906	0.9909	0.9911	0.9913	0.9916
2.4	0.9918	0.9920	0.9922	0.9925	0.9927	0.9929	0.9931	0.9932	0.9934	0.9936
2.5	0.9938	0.9940	0.9941	0.9943	0.9945	0.9946	0.9948	0.9949	0.9951	0.9952
2.6	0.9953	0.9955	0.9956	0.9957	0.9959	0.9960	0.9961	0.9962	0.9963	0.9964
2.7	0.9965	0.9966	0.9967	0.9968	0.9969	0.9970	0.9971	0.9972	0.9973	0.9974
2.8	0.9974	0.9975	0.9976	0.9977	0.9977	0.9978	0.9979	0.9980	0.9980	0.9981
2.9	0.9981	0.9982	0.9983	0.9983	0.9984	0.9984	0.9985	0.9985	0.9986	0.9986
3	0.9987	0.9987	0.9987	0.9988	0.9988	0.9989	0.9989	0.9989	0.9990	0.9990
3.1	0.9990	0.9991	0.9991	0.9991	0.9992	0.9992	0.9992	0.9992	0.9993	0.9993
3.2	0.9993	0.9993	0.9994	0.9994	0.9994	0.9994	0.9994	0.9995	0.9995	0.9995

2.4.3.1   The Memoryless Property of the Exponential Distribution

A very special property of the continuous exponential distribution is that it has no memory. Let the waiting time of ε process being modeled by means of an exponential distribution of a random variable X. If it is known that the waiting time is greater than a certain value t_1, then the probability of the waiting time less than $t_1 + t_2$ equals to the probability of the waiting time of t_2. Such as;

$$P\left(X < t_1 + t_2 \mid X > t_1\right) = P\left(X < t_2\right).$$

2.4.3.2   Proof

According to Bayes' theory for the conditional probability;

$$P\left(X < t_1 + t_2 \mid X > t_1\right) = \frac{P\left(t_1 < X < t_1 + t_2\right)}{P\left(X > t_1\right)}$$

$$= \frac{F_X\left(t_1 + t_2\right) - F_X\left(t_1\right)}{1 - F_X\left(t_1\right)}$$

since $F_X(x) = 1 - e^{-\lambda x}$

$$P\left(X < t_1 + t_2 \mid X > t_1\right) = \frac{1 - e^{-\lambda(t_1 + t_2)} - \left(1 - e^{-\lambda t_1}\right)}{1 - \left(1 - e^{-\lambda t_1}\right)}$$

$$= 1 - e^{-t_2} = F_X\left(t_2\right) = P\left(X < t_2\right).$$

Example 2.6 The probability density function of a random variable X is given as;

$$f_X(x) = \begin{cases} \frac{1}{2} e^{\frac{-1}{2}(x-3)} & x \geq 3 \\ 0 & \text{elsewhere} \end{cases}$$

Calculate the probabilities that;

(a) $P(X < 4)$
(b) $P(X < 5 \mid x > 4)$
(c) Compare the results obtained from part (a) and part (b).

Solution 2.6 (a)

$$P(X < 4) = \int_{-\infty}^{4} f_X(x)\mathrm{d}x = \int_{3}^{4} \frac{1}{2} e^{\frac{-1}{2}(x-3)}\mathrm{d}x$$

$$= - e^{\frac{-1}{2}(x-3)}\Big|_{3}^{4} = 1 - e^{\frac{-1}{2}}$$

(b)

$$P(X < 5 \mid x > 4) = \frac{P(4 < X < 5)}{P(X > 4)} = \frac{\int_4^5 \frac{1}{2} e^{\frac{-1}{2}(x-3)} dx}{1 - \int_3^4 \frac{1}{2} e^{\frac{-1}{2}(x-3)} dx}$$

$$= \frac{e^{\frac{-1}{2}} - e^{-1}}{e^{\frac{-1}{2}}} = 1 - e^{\frac{-1}{2}}$$

(c) As expected, according to memoryless property of the exponential distribution,
the results obtained from part (a) and part (b) are the same.

2.4.4 *Gamma Distribution*

The probability density function of a gamma distribution is;

$$f_X(x) = \begin{cases} \frac{\lambda^\alpha x^{\alpha-1} e^{-\lambda x}}{\Gamma(\alpha)} & x > 0 \\ 0 & \text{elsewhere} \end{cases}$$

where $\lambda > 0$, and $\alpha > 0$. $\Gamma(\alpha)$ is called Gamma function that equals;

$$\Gamma(\alpha) = \int_0^\infty x^{\alpha-1} e^{-x} dx$$

The expected value of a gamma distribution is;

$$\mu_X = E[X] = \frac{\alpha}{\lambda}$$

and the variance of a gamma distribution is;

$$\text{Var}[X] = \sigma_X^2 = \frac{\alpha}{\lambda^2}$$

The gamma distribution becomes the exponential distribution when

$$\alpha = 1$$

An illustration of the Gamma distribution with various parameters are shown in
Fig. 2.11.

A detailed table for the continuous distributions is supplied in Appendix section.

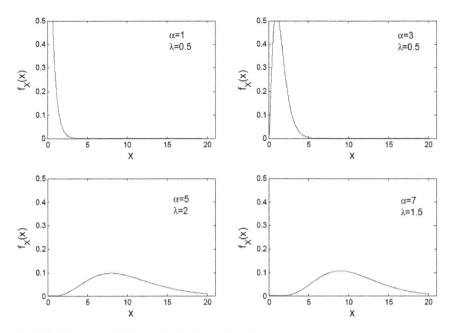

Fig. 2.11 Illustration of Gamma distribution with various parameters

2.5 Problems

2.1 Let the random variable X have the following probability density function;

$$f_X(x) = \begin{cases} c & 3 \le x \le 5 \\ 0 & \text{elsewhere} \end{cases}$$

(a) Find c such that $f_X(x)$ is a valid probability density function.
(b) Find $Fx(x)$, and plot it.
(c) Find the probabilities of $P(X < 1, P(X \ge 2, P(X = 4))$.

2.2 Consider the following probability density function of X

$$f_X(x) = \begin{cases} \frac{c}{2} & \text{for } 0 \le x \le 1 \\ c & 1 \le x \le 2 \\ 0 & \text{elsewhere} \end{cases}$$

(a) Find c such that $f_X(x)$ is a valid probability density function.
(b) Find $F_X(x)$, and plot it.
(c) Find the probabilities of $P(0 < X < 1, P(X \ge 1.5, P(X = 1.7)$.

2.3 The probability density function of a random variable X is given as;

$$f_X(x) = \begin{cases} Ae^{-(x+2)} & \text{for } x \geq -2 \\ 0 & \text{elsewhere} \end{cases}$$

(a) Find A such that $f_X(x)$ is a valid probability density function.
(b) Find the probabilities of $P(-1 < X < 0)$, $P(X \geq 5)$, $P(X = 0)$.
(c) Find $F_X(x)$.
(d) Repeat part (b) using part (c).

2.4 The failure time of a brand new car is exponentially distributed with the expected value of 5 years.

(a) What is the probability that the car fails within the first three years period?
(b) If it is known that the car failed within the first 8 years period, what is the probability that the failure occurred between the 3rd year and the 5th year?

2.5 The random variable X has the following probability density function;

$$f_X(x) = \frac{\frac{\beta}{\pi}}{\beta^2 + (x - \alpha)^2}$$

where $\beta > 0$, and $-\infty < \alpha < \infty$. It is known as Cauchy distribution.

(a) Derive the probability distribution function for $\alpha = -2$, and $\beta = 1$
(b) Find the probabilities that $P(X < -2$, $P(-1 < X \leq 1)$, $P(X = 3)$.

2.6 The Laplace distribution has a probability density function;

$$f_X(x) = \frac{\beta}{2} e^{-\beta|x - \alpha|}$$

where $\beta > 0$, and $-\infty < \alpha < \infty$. It is known as Cauchy distribution.

(a) Derive the probability distribution function in general form.
(b) Calculate the probabilities that $P(X > 2)$, $P(-1 < X \leq 2)$ for $\alpha = 2$, and $\beta = 4$
(c) Find μ_X and σ_X.

2.7 A professor may start a quiz in the first half an hour of the lecture duration. Let the starting time exponentially distributed as;

$$f_X(x) = \begin{cases} \left(ke^{-0.1x}\right) & \text{for } 0 < x \leq 30 \\ 0 & \text{elsewhere} \end{cases}$$

(a) Find k in order to have a proper the probability density function.
(b) What is the probability that the quiz starts within the first 10 min?

(c) Calculate the probability that the quiz starts after the 15 min.

2.8 Let X is a continuous random variable with the probability distribution function $F_X(x)$. An event E is described as $E = \{x > a\}$, where a is a finite real number. Derive the conditional distribution function $F_X(x \mid E)$ in terms of $F_X(x)$.

2.9 Repeat the Problem 2.8 if X is exponentially distributed with $\mu_X = 5$ and $a = 7$.

2.10 The general form of Rayleigh probability distribution function is

$$F_X(x) = \begin{cases} \left(1 - e^{-\frac{(x-\alpha)^2}{\beta}}\right) & \text{for } x \geq \alpha \\ 0 & \text{elsewhere} \end{cases}$$

(a) Derive $f_X(x)$ where $\beta = 2$, and $\alpha = -4$
(b) Calculate the probabilities that $P(X \leq -3, P(X \geq -5, P(-7 < X \leq -2), P(X > 4)$.

2.11 The probability density function of the time-to-failure of a particular color TV model is given as;

$$f_X(x) = \frac{1}{8} e^{\frac{-x}{8}} u(x)$$

What is the expected life time of a product purchased?

2.12 Suppose we choose a capacitor with capacitance of C from the batch of capacitors normally distributed with parameters $\mu = 100$ Farad, $\sigma = 25$ Farad.

(a) What is the probability that C will have a value between 90 and 110 Farad?
(b) What is the probability that C will have a value between 50 and 90 Farad?
(c) If it is known that the chosen capacitor has its value >70 Farad; what is the probability that the capacitance value lies between 82 and 96 Farad?

2.13 For the given probability density function

$$f_X(x) = \begin{cases} A \cos\left(\frac{\pi}{4}x\right) & \text{if } -2 < x \leq 2 \\ 0 & \text{elsewhere} \end{cases}$$

(a) Find A such that $f_X(x)$ is a valid pdf
(b) Find the probability distribution function $F(x)$
(c) Compute the following probabilities using $F(x)$

   (i) $P\{-1 < x \leq 0\}$
   (ii) $P\{x > 0\}$.

2.14 Find the mean and variance of the following probability density function;

$$f_x(x) = \frac{1}{2} e^{|x|}.$$

Chapter 3
Discrete Random Variables

In many applications, the random variable X has to be defined in discrete form as a result of the problem in which it is used. For instance, the flipping a fair coin experiment has only two possible outcomes, namely "head" and "tail". Let $X(head) = 1$ and $X(\text{tail}) = 0$, then it is written that $P(X = 1) = \frac{1}{2}$, and $P(X = 0) = \frac{1}{2}$.

Throwing a fair dice, number of people waiting n a banking queue, number of defected data packets in an information system can be encountered for further examples having discrete forms.

Similar to the continuous case, the probability functions are employed to represent a discrete random variable.

3.1 The Probability Distribution Function (PDF)

Discrete random variables have stair-case probability distribution functions having combinations of jumps at certain values and remaining constant at rest of the values. An illustrative graph is shown in Fig. 3.1. The relationship between $F_X(x)$ and the probability operation is;

$$F_X(a) = P(X \leq a) \quad \text{for } -\infty < a < \infty$$

Properties of $F_X(x)$
Considering the axioms of the probability theory introduced in Chap. 1, the properties of a discrete $F_X(x)$ are enumerated as;

© The Author(s), under exclusive license to Springer Nature Switzerland AG 2022
M. Catak et al., *Probability and Random Variables for Electrical Engineering*,
Studies in Systems, Decision and Control 390,
https://doi.org/10.1007/978-3-030-82922-3_3

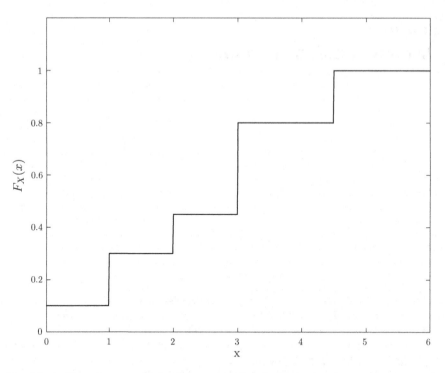

Fig. 3.1 An illustrative example of a discrete probability distribution function

1. $0 \le F_X(x) \le 1$ for all x values.
2. $F_X(x)$ is a nondecreasing function, since it is a cumulative function,

$$F_X(x_1) \le F_X(x_2) \quad \text{where} \quad x_1 < x_2$$

3. $\lim_{x \to \infty} F_X(x) = 1$
4. $\lim_{x \to -\infty} F_X(x) = C$

3.2   The Probability Density Function (PDF)

The probability density function called probability mass function has nonzero values at certain points where the jumps take place at the probability distribution function. In another words,

$$f_X(x) = \lim_{\epsilon \to 0} F_X(x + \epsilon) - F_X(x)$$

Properties of $f_X(x)$

1. Since $F_X(x)$ is a non-decreasing function, $f_X(x)$ must be greater than or equal to 0.

$$f_X(x) \geq 0$$

2. Sum of $f_X(x)$ over its domain must be 1.

$$\sum f_X(x_i) = 1$$

3. $F_X(x)$ is calculated as

$$F_X(x) = P(X \leq x) = \sum_{x_i \leq x} f_X(x_i)$$

4.

$$P(X = a) = f_X(a)$$

The 4th property is one of the main differences between the continuous and the discrete random variables. As we discussed in Chap. 2, for the continuous case the following equalities are valid;

$$P(a < X \leq b) = P(a < X < b) = P(a \leq X \leq b) = P(a \leq X < b)$$
$$= F_X(b) - F_X(a)$$

However, those equalities may or may not hold for the discrete case depending on its probability density function.

For the sake of well-understanding of discrete probability density functions, and consequently the probability distribution functions, the Dirac delta function, and the unit step function have to be introduced.

The Dirac Delta Function

The Dirac delta function $\delta(x)$, which is called unit impulse function as well in the literature, is defined as;

$$\delta(x - k) = \begin{cases} 1 & \text{for } x = k \\ 0 & \text{elsewhere} \end{cases}$$

and

$$\int_{-\infty}^{\infty} \delta(x)\mathrm{d}x = 1$$

For $f(x)$ is any continuous function at $x = k$

$$\int_{-\infty}^{\infty} f(x)\delta(x-k)dx = f(k)$$

The Unit Step Function

The unit step function is defined as;

$$u(x-k) = \begin{cases} 1 & \text{for } x \geq k \\ 0 & \text{for } x < k \end{cases}$$

The relationship between the Dirac delta function, and the unit step function is;

$$\int_{-\infty}^{x} \delta(\tau-k)d\tau = u(x-k)$$

$$\frac{du(x)}{dx} = \delta(x)$$

Example 3.1 Let a random variable X has the following probability density function;

$$f_X(x) = 0.2\delta(x-1) + 0.3\delta(x) + 0.1\delta(x+1) + 0.3\delta(x+2) + 0.16(x+3)$$

Find the probabilities that $P(X \leq 1), P(X < 1), P(X = 0)$ $P(X = 4)$, $P(X > 5)$.

Solution 3.1 The probability density function $f_X(x)$ is illustrated in Fig. 3.2.
The corresponding probability distribution function is shown in Fig. 3.3.
Thereafter the probabilities are calculated as;

$$P(X \leq 1) = F_X(1) = 1$$
$$P(X < 1) = \lim_{x \to 1^-} F_X(x) = 0.8$$
$$P(X = 0) = F_X\left(0^+\right) - F_X\left(0^-\right) = 0.3$$
$$P(X = 4) = F_X\left(4^+\right) - F_X\left(4^-\right) = 0$$
$$P(X \geq 5) = 1 - F_X(5) = 0$$

Example 3.2 Let the probability density function be defined as;

$$f_X(x) = \sum_{x_i=1}^{\infty} \left(\frac{1}{2}\right)^{x_i} \delta(x - x_i)$$

(a) Show that $f_X(x)$ is a valid probability density function.
(b) Find the probabilities of $P(2 < X \leq 4, P(X > 3, P(X = 5)$.

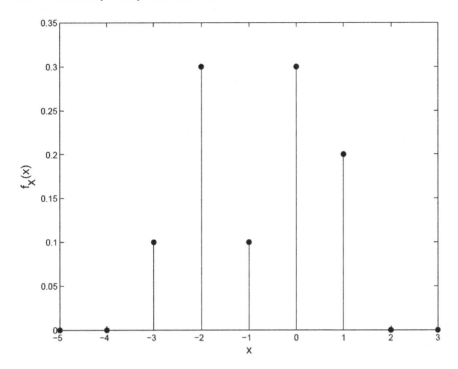

Fig. 3.2 Discrete probability density function (Example 3.1)

Solution 3.2 (a) A proper probability density function satisfies the following conditions;

$$f_X(x) \geq 0$$
$$\sum f_X(x_i) = 1$$

For all values of $x f_X(x)$ is positive. It is known that

$$\sum_{i=1}^{\infty} r^i = \frac{r}{1-r}$$

where $|r| < 1$. Then,

$$\sum f_X(x_i) = \sum_{x_i=1}^{\infty} \left(\frac{1}{2}\right)^{x_i} = \frac{\frac{1}{2}}{1-\frac{1}{2}} = 1$$

Therefore, $f_X(x)$ is a proper probability density function.
(b) Find the probabilities of $P(2 < X \leq 4, P(X > 3, P(X = 5))$

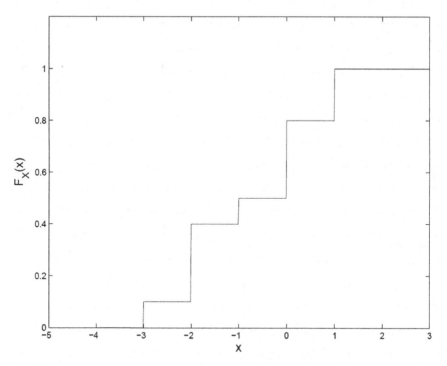

Fig. 3.3 Discrete probability distribution function (Example 3.1)

$$P(2 < X \le 4) = \left(\frac{1}{2}\right)^3 + \left(\frac{1}{2}\right)^4$$
$$P(X > 3) = 1 - P(X \le 3)$$
$$= 1 - \frac{1}{2} - \left(\frac{1}{2}\right)^2 - \left(\frac{1}{2}\right)^3$$
$$P(X = 5) = \left(\frac{1}{2}\right)^5.$$

3.3 Expected Value and Variance

The expected value and the variance of a discrete random variable can be calculated by means of its probability density function;

$$E[X] = \mu_X = \sum x_i f_X(x_i)$$
$$\text{Var}[X] = \sigma_X^2 = \sum (x_i - \mu_X)^2 f_X(x_i).$$

Example 3.3 Find the expected value and the variance of the probability distribution given in Example 3.1.

Solution 3.3 Since the expected value can be calculated as;

$$E[X] = \mu_X = \sum x_i f_X(x_i)$$
$$= (-3 \times 0.1) + (-2 \times 0.3) + (-1 \times 0.1) + (0 \times 0.3) + (1 \times 0.2)$$
$$= -0.8$$

and the corresponding variance is derived as;

$$Var[X] = \sigma_X^2 = \sum (x_i - \mu_X)^2 f_X(x_i)$$
$$= (-3 + 0.8)^2 \times 0.1 + (-2 + 0.8)^2 \times 0.3$$
$$+ (-1 + 0.8)^2 \times 0.1 + (0 + 0.8)^2 \times 0.3 + (1 + 0.8)^2 \times 0.2$$
$$= 1.76.$$

3.4 Common Discrete Probability Distribution Functions

3.4.1 Uniform Distribution

A discrete random variable X is called uniformly distributed if X has equally likely probabilities at a finite number of x. The corresponding probability density function is expressed as (Fig. 3.4);

$$f_X(x) = \sum_{i=1}^{n} \frac{1}{n} \delta(x - x_i)$$

The probabilities of the outcomes of a fair dice experiment is a good example for a discrete uniform distribution.

Let the uniform distribution is defined in the interval $[a, b]$. That means $x_1 = a$, and $x_n = b$. Then, the number of the points where X has nonzero probability is $n = b - a + 1$ The expected value of a uniform distribution is calculated as;

$$E[X] = \sum_{i=a}^{b} i \frac{1}{b-a+1} = \frac{1}{b-a+1} \sum_{i=a}^{b} i = \frac{1}{b-a+1} \left(\sum_{i=1}^{b} i - \sum_{i=1}^{a} i \right)$$
$$= \frac{b(b+1) - a(a-1)}{2(b-a+1)} = \frac{a+b}{2}$$

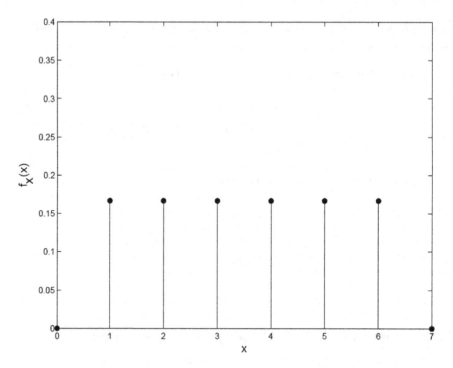

Fig. 3.4 Discrete probability distribution function of a fair dice experiment

Similarly, the variance of the uniform discrete distribution is founded as;

$$\text{Var}[X] = \frac{(b - a + 1)^2 - 1}{12}.$$

3.4.2 Bernoulli Distribution

Bernoulli distribution is used to model such processes having only two possible outcomes. For instance "pass" or "fail", "win" or "loose", etc. The probability density function is given as (Fig. 3.5);

$$f_X(x) = \begin{cases} p, & x = 0 \\ q, & x = 1 \end{cases}$$

where $p \geq 0, q \geq 0$, and $p + q = 1$ It can be represented by the Dirac delta function,

$$f_X(x) = p\delta(x) + q\delta(x - 1)$$

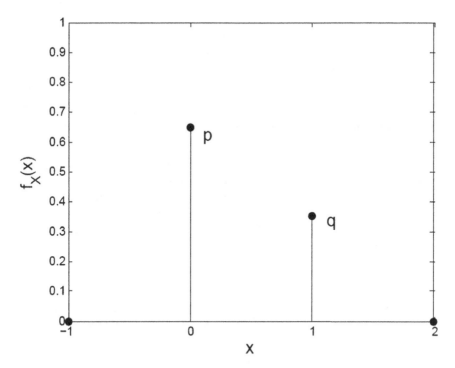

Fig. 3.5 Bernoulli probability density function

The expected value of a Bernoulli distribution is;

$$E[X] = p$$

and its variance is;

$$\text{Var}[X] = pq = p(1 - p).$$

3.4.3 Binomial Distribution

The probability density function of a binomial distribution is

$$f_X(x) = \sum_{x_i} \binom{n}{x} p^x (1 - p)^{n-x} \delta (x - x_i)$$

Illustrative examples for the binomial distribution are shown in Fig. 3.7 for various n, and p values.

The expected value and the variance of a binomial distribution are derived as;

$$E[X] = np$$
$$\text{Var}[X] = npq = np(1 - p).$$

3.4.4 Geometric Distribution

Geometric distribution is a kind of sequential realization of a Bernoulli distribution. Let consider that p is the probability of being "success" and $q = 1 - p$ is the probability of being "fail" governed by a Bernoulli distribution. The number of trials until the first success obtained is assigned to the random variable X, then X is called a geometrically distributed random variable with the parameter of p. The corresponding probability density function is;

$$f_X(x) = \sum_{x_i}(1 - p)^{x-1} p \delta (x - x_i), \quad x_i = 1, 2, \ldots$$

The expected value and the variance of a Geometric distribution are;

$$E[X] = \frac{1}{p}$$
$$\text{Var}[X] = \frac{1 - p}{p^2}.$$

3.4.5 The Memoryless Property of the Geometric Distribution

A system that has been modeled by means of geometric distribution has no memory. In other words, the probability that the first success obtained less than $n_1 + n_2$ trials with the condition that the first success has not taken place within the first n_1 trials equals to the probability that the first success obtained less than n_2 trials, such that (Fig. 3.6);

$$P(X \leq n_1 + n_2 \mid X > n_1) = P(X \leq n_2).$$

Proof According to Bayes' theory for the conditional probability;

$$P(X \leq n_1 + n_2 \mid X > n_1) = \frac{P(n_1 < X \leq n_1 + n_2)}{P(X > n_1)}$$
$$= \frac{\sum_{i=n_1+1}^{n_1+n_2}(1 - p)^{i-1} p}{1 - \sum_{i=1}^{n_1}(1 - p)^{i-1} p}$$

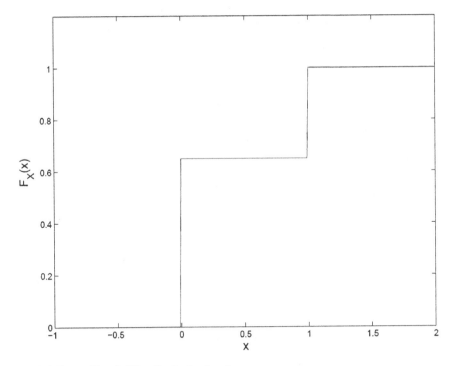

Fig. 3.6 Bernoulli probability distribution function

since;

$$\sum_{k=0}^{n}(z)^k = \frac{1 - z^{n+1}}{1 - z}$$

$$P\left(X \le n_1 + n_2 \mid X > n_1\right) = \frac{\sum_{i=1}^{n_1}(1 - p)^i - \sum_{i1}^{n_1+n_2}(1 - p)^i}{\sum_{i=1}^{n_1}(1 - p)^i}$$

$$= \sum_{i=1}^{n_2}(1 - p)^i = P\left(X \le n_2\right).$$

3.4.6 Poisson Distribution

The probability density function of a Poisson distribution can be written as;

$$f_X(x) = e^{-\lambda} \sum_{k=0}^{\infty} \frac{\lambda^k}{k!} \delta(x - k)$$

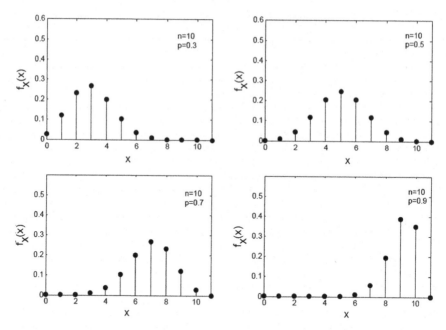

Fig. 3.7 Binomial probability distribution function

The corresponding probability distribution function is (Fig. 3.8);

$$F_X(x) = e^{-\lambda} \sum_{k=0}^{\infty} \frac{\lambda^k}{k!} u(x - k)$$

The expected value and the variance of a Poisson distribution have the same value of the parameter λ (Fig. 3.8).

$$E[X] = \lambda$$
$$\text{Var}[X] = \lambda.$$

Example 3.4 The waiting time of a call center is a random variable depending on a Poisson distribution with the mean of 4 min. Find the probabilities that

(a) $P(X = 0)$
(b) $P(X > 2)$
(c) $P(X \leq 3)$
(d) $P(X \leq 8 \mid X > 5)$.

Solution 3.4

(a)

$$P(X = 0) = f_X(0) = e^{-4}$$

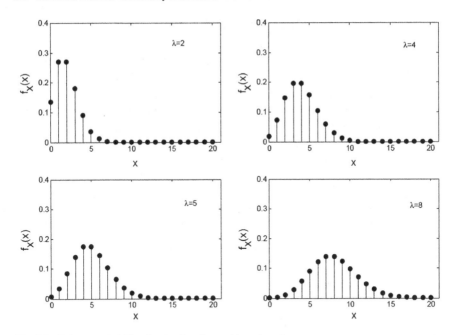

Fig. 3.8 Poisson probability density functions with various λ

(b)

$$P(X > 2) = 1 - P(X \leq 2) = 1 - \left(e^{-4} + e^{-4}4 + \frac{e^{-4}4^2}{2!} \right)$$

(c)

$$P(X \leq 3) = e^{-4} + e^{-4}4 + \frac{e^{-4}4^2}{2!} + \frac{e^{-4}4^3}{3!}$$

(d)

$$
\begin{aligned}
P(X \leq 8 \mid X > 5) &= \frac{P(X \leq 8) \cap P(X > 5)}{P(X > 5)} \\
&= \frac{P(X = 6, 7, 8)}{P(X > 5)} \\
&= \frac{\frac{e^{-4}4^6}{6!} + \frac{e^{-4}4^7}{7!} + \frac{e^{-4}4^8}{8!}}{1 - \left(e^{-4} + e^{-4}4 + \frac{e^{-4}4^2}{2!} + \frac{e^{-4}4^3}{3!} + \frac{e^{-4}4^4}{4!} + \frac{e^{-4}4^5}{5!} \right)}.
\end{aligned}
$$

3.5 The Mixed Probability Distributions

In practical applications, a random variable may consists of both discrete and continuous probability density functions in order to give better representation of a physical system.

The mixed probability density function obeys all the properties of a proper probability density function. Let the mixed probability density function be

$$f_{X,m}(x) = f_{X,c}(x) + f_{X,d}(x)$$

where, $f_{X,m}(x)$ is the mixed pdf, $f_{X,c}(x)$ is the continuous part of the mixed pdf, and $f_{X,d}(x)$ is the discrete part of the mixed pdf. If;

$$\int_{-\infty}^{\infty} f_{X,c}(x)dx = A \quad \text{where} \quad 0 < A < 1$$

then, according to the fundamental probability law;

$$\sum f_{X,d}(x) = 1 - A$$

Thereafter, the expected value of a mixed probability distribution is calculated as a weighted average of its parts;

$$E[X] = (1 - A) \sum_{x_i} x_i f_{X,d} + A \int_{-\infty}^{\infty} x f_{X,c}(x)dx$$

The variance of a mixed probability distribution is;

$$E[X] = (1 - A) \sum_{x_i} (x_i - E[X])^2 f_{X,d} + A \int_{-\infty}^{\infty} (x - E[X])^2 f_{X,c}(x)dx.$$

Example 3.5 The random variable X has the probability density function illustrated in Fig. 3.9

(a) Find c such that $f_X(x)$ is a proper probability density function
(b) Find and plot $F_X(x)$
(c) Calculate the probabilities of $P(0 < x \leq 1)$, $P(X = 0.5)$, $P(X = 0.6)$, $P(X > 1)$.

Solution 3.5 (a) The continuous part of the mixed pdf is;

$$f_{X,c}(x) = \begin{cases} cx, & 0 < x \leq 1 \\ -cx + 1, & 1 < x \leq 2 \\ 0, & \text{elsewhere} \end{cases}$$

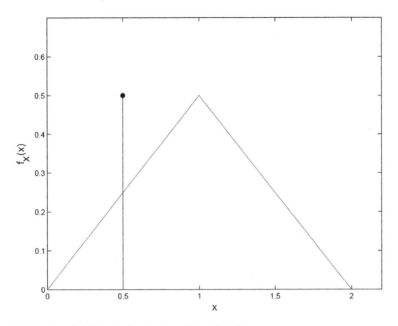

Fig. 3.9 Mixed probability density function (Example 3.5)

and the discrete part of the mixed pdf;

$$f_{X,d}(x) = c\delta(x - 0.5)$$

Thereby,

$$\int_0^1 cx\,dx + \int_1^2 -cx\,dx + c = 1$$

then, one can easily find that $c = 0.5$

(b) Case $I : x < 0$

$$F_X(x) = \int_{-\infty}^x f_X(\tau)d\tau = \int_{-\infty}^x 0\,d\tau = 0$$

Case $II : 0 \le x < 0.5$

$$F_X(x) = \int_{-\infty}^x f_X(\tau)d\tau$$

$$= \int_0^x 0.5\,d\tau$$

$$= \left(\frac{\tau^2}{4}\right)_0^x$$

$$= \frac{x^2}{4}$$

Case $III : 0.5 \leq x < 1$

$$F_X(x) = \int_{-\infty}^{x} f_X(\tau)d\tau$$

$$= \frac{1}{16} + \frac{1}{2} + \int_{0.5}^{x} \frac{\tau}{2}d\tau$$

$$= \frac{9}{16} + \left(\frac{\tau^2}{4}\right)_{0.5}^{x}$$

$$= 0.5 + \frac{x^2}{4}$$

Case $IV : 1 \leq x < 2$

$$F_X(x) = \int_{-\infty}^{x} f_X(\tau)d\tau = \frac{12}{16} + \int_{1}^{x} \left(\frac{-\tau}{2} + 1\right) d\tau$$

$$= \frac{12}{16} + \left(\frac{-\tau^2}{4} + \tau\right)_{1}^{x} = x - \frac{x^2}{4}$$

Case $V : x \geq 2$

$$F_X(x) = \int_{-\infty}^{x} f_X(\tau)d\tau = 1$$

then,

$$F_X(x) = \begin{cases} 0 & x < 0 \\ \frac{x^2}{4} & 0 \leq x < 0.5 \\ 0.5 + \frac{x^2}{4} & 0.5 \leq x < 1 \\ x - \frac{x^2}{4} & 1 \leq x < 2 \\ 1 & x \geq 2 \end{cases}$$

(c)

$$P(0 < x \leq 1) = F_X(1) - F_X(0) = 0.75$$
$$P(X = 0.5) = F_X(0.5) - F_X\left(0.5^-\right) = f_{X,d}(0.5) = 0.5$$
$$P(X = 0.6) = 0$$
$$P(X > 1) = 1 - F_X(1) = 0.25.$$

3.6 Problems

3.1 The probability density function of the number of cars in a car park is defined using Poisson distribution as;

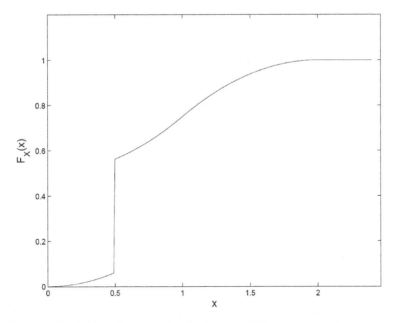

Fig. 3.10 Mixed probability distribution function (Example 3.5)

$$f(x) = \begin{cases} e^{-15}\frac{15^x}{x!} & \text{if } x = 0, 1, 2, , , \\ 0 & \text{elsewhere} \end{cases}$$

(a) Find the probability distribution function $F(x)$
(b) What is the probability of number of cars is being between 5 and 10 (5 and 10 are included)

3.2 X has the discrete values $\{-1, -0.5, 0.7, 1.5, 3\}$. The corresponding probabilities are assumed to be $\{0.1, 0.2, 0.1, 0.4, 0.2\}$

(a) Plot $f_X(x)$
(b) Plot $F_X(x)$
(c) Compute the probabilities of $P\{0 < X \leq 1\}$, $P\{X = 0.7\}$
(d) Find μ_X.

3.3 A random variable X is geometrically distributed with $p = 0.6$. Find the following probabilities that (Fig. 3.10)

(a) $P(X = 2)$
(b) $P(X \geq 3)$
(c) $P(X \leq 4)$
(d) $P(X \leq 7 \mid X > 3)$.

3.4 The discrete random variable X has the probability function

$$f_X(x) = \begin{cases} 3cx & x = 1, 3, 5 \\ c(x-2) \ x = 7 \\ 0 & \text{elsewhere} \end{cases}$$

(a) Find c in order to have a proper the probability density function.
(b) Calculate the probabilities that $P(X \leq 2, P(X \geq 4, P(0 < X \leq 3), P(X > 4)$
(c) Find the expected value of the distribution, $E[X]$.
(d) Calculate the variance of the distribution, $Var[X]$.

3.5 Let X be a discrete random variable with the following probability density function

$$f_X(x) = \begin{cases} 0.04 \ \text{for } x = 0 \\ 0.08 \ \text{for } x = 2 \\ 0.12 \ \text{for } x = 4 \\ 0.16 \ \text{for } x = 6 \\ 0.20 \ \text{for } x = 8 \\ 0.16 \ \text{for } x = 10 \\ 0.12 \ \text{for } x = 12 \\ 0.08 \ \text{for } x = 10 \\ 0.04 \ \text{for } x = 12 \end{cases}$$

(a) Find $E[X]$.
(b) Find Var$[X]$ and σ_X.
(c) Calculate the probabilities of $P(X \leq 5)$, $P(X \leq 5 \mid X > 1)$.

3.6 A random variable X has the following probability distribution function;

$$F_X(x) = \begin{cases} 0 & \text{for } x < 0 \\ \frac{1}{4} & \text{for } 0 \leq x < 1 \\ \frac{2}{5} & \text{for } 1 \leq x < 3 \\ \frac{7}{10} & \text{for } 3 \leq x < 5 \\ 1 & \text{for } x \geq 5 \end{cases}$$

Determine the probability density function of X, $f_X(x)$.

3.7 The random variable X has the probability density function;

$$f_X(x) = 0.25\delta(x+1) + 0.25\delta(x) + g(x)$$

where

$$g(x) = \frac{1}{4}x + \frac{1}{4} \quad \text{if} -1 < x < 1$$

(a) Show that $f_X(x)$ is a proper probability density function
(b) Find and plot $F_X(x)$
(c) Calculate the probabilities of $P(0 \le x \le 1)$, $P(X = -1)$, $P(X = 0)$, $P(X > 1)$.

Chapter 4
Multiple Random Variables

In the previous chapters single random variables have been discussed in continuous, discrete, and mixed forms based on the probability functions. It is also possible to define multiple random variables in a probability space. In that case, the probability functions are defined using n random variables, where $n \geq 2$. The probability functions in $\mathbb{R}$ (i.e. a line) for the single random variables, they works on $\mathbb{R}^2$ (i.e. a plane) for the two-random variables, and so on.

4.1   The Joint Probability Distribution Function

For the sake of simplicity and to gain better understanding, let discuss a two-dimensional form at first. X and Y are two random variables that are defined on the same probability space. The joint probability distribution is given as;

$$F_{XY}(x, y) = P(X \leq x, Y \leq y)$$

Properties of the Joint PDF $F_{XY}(x, y)$

1. Since $F_{XY}(x, y)$ is a non-decreasing function, then $F_{XY}(x, y)$ must be greater than or equal to 0.

$$F_{XY}(x, y) \geq 0$$

M. Catak et al., *Probability and Random Variables for Electrical Engineering*,
Studies in Systems, Decision and Control 390,
https://doi.org/10.1007/978-3-030-82922-3_4

2. $0 \leq F_{XY}(x, y) \leq 1$
3. $F_{XY}(-\infty, y) = F_{XY}(x - \infty) = F_{XY}(-\infty, -\infty) = 0$
4. $F_{XY}(\infty, \infty) = 1$.

4.2 The Joint Probability Density Function

For the continuous random variables, the joint probability density function, $f_{XY}(x, y)$, is calculated by taking the partial derivatives of $F_{XY}(x, y)$ with respect to x, and y.

On the other hand, for the discrete random variables the probability density function is defined as;

$$f_{XY}(x, y) = P(x = x, Y = y)$$

Properties of the Joint PDF $f_{XY}(x, y)$
1. $f_{XY}(x, y) \geq 0$
2. $\int_{-\infty}^{\infty} \int_{-\infty}^{\infty} f_{XY}(x, y)dxdy = 1$ continuous case
   $\sum_x \sum_y f_{XY}(x, y) = 1$ discrete case
3. The joint probability distribution function, $Fxy(x, y)$, is calculated by means of the Joint probability density function as;
   $F_{XY}(x, y) = \int_{-\infty}^{x} \int_{-\infty}^{y} f_{XY}(\tau, \xi)d\tau d\xi$ continuous case
   $F_{XY}(x, y) = \sum_x \sum_y f_{XY}(x, y)$ discrete case
4. $P(x_1 < X \leq x_2, y_1 < Y \leq y_2) = \int_{x_1}^{x_2} \int_{y_1}^{y_2} f_{XY}(x, y)dxdy$ continuous case
   $P(x_1 < X \leq x_2, y_1 < Y \leq y_2) = \sum_{x=x_1}^{x_2} \sum_{y=y_1}^{y_2} f_{XY}(x, y)$ discrete case
5. $f_{XY}(x, y) = \frac{\partial^2 F_{XY}(x,y)}{\partial x \partial y}$.

4.3 Marginal Probability Functions

Although, the joint probability functions characterize a process modeled by means of multiple random variables, it is also important to derive individual effects of each single random variables on such a process. The individual probability functions of a single random variable are called marginal probability functions.

4.3.1   *Marginal Probability Distribution Function*

Let consider a joint probability distribution function of two random variables, $F_{XY}(x, y)$. The marginal probability distribution function of the random variable Y is calculated as;

$$F_Y(y) = F_{XY}(\infty, y) \quad \text{continuous case}$$

$$F_Y(y) = \sum_x F_{XY}(x, y) \quad \text{discrete case}$$

Similarly, the marginal probability distribution function of the random variable Y is derived as;

$$F_X(x) = F_{XY}(x, \infty) \quad \text{continuous case}$$

$$F_X(x) = \sum_y F_{XY}(x, y) \quad \text{discrete case}$$

4.3.2   *Marginal Probability Density Function*

The marginal probability density function of the random variable Y is;

$$f_Y(y) = \int_{-\infty}^{\infty} f_{XY}(x, y) \mathrm{d}x \quad \text{continuous case}$$

$$f_Y(y) = \sum_x f_{XY}(x, y) \quad \text{discrete case}$$

The marginal probability density function of the random variable X is;

$$f_X(x) = \int_{-\infty}^{\infty} f_{XY}(x, y) \mathrm{d}y, \quad \text{continuous case} .$$

$$f_X(x) = \sum_y f_{XY}(x, y), \quad \text{discrete case}$$

4.4   Conditional Probability and Statistical Independence

The conditional probability of the event A under the condition that the event B has already occurred is described as;

$$P(A|B) = \frac{P(A \cap B)}{P(B)}$$

where $P(A) \neq 0$.

If A and B are mutually exclusive, that is;

$$P(A \cap B) = P(A)P(B)$$

then, the conditional probability can be written as;

$$P(A|B) = \frac{P(A)P(B)}{P(B)}$$

In the same manner, if X and Y are statistically independent random variables, the condition about the random variable Y does not affect the probability that is related to the random variable X. Therefore, the corresponding marginal probability density function of X can be calculated as;

$$f_X(x) = \frac{f_{XY}(x, y)}{f_Y(y)}$$

where $f_Y(y) \neq 0$.

4.5   Functions of Random Variables

4.5.1 $Y = aX + b$, $a > 0$, *Continuous Random Variables*

Let X be a random variable and random variable Y is defined as

$$Y = aX + b$$

where $a > 0$

The probability distribution function of the random variable Y can be expressed using the definition of the probability function such that;

$$\begin{aligned}
F_Y(y) &= P(Y \leq y) \\
&= P(aX + b \leq y) \\
&= P\left(X \leq \frac{y - b}{a}\right) \\
&= F_X\left(\frac{y - b}{a}\right)
\end{aligned}$$

The probability density function of Y is defined as,

$$f_Y(y) = \frac{d\,F_Y(y)}{dy} = \frac{d\,F_X\left(\frac{y-b}{a}\right)}{dy}$$

$$= \frac{1}{a} f_X\left(\frac{y-b}{a}\right).$$

Example 4.1 Consider a random variable X having the following probability density function

$$f_X(x) = \begin{cases} \frac{1}{2} & 0 \le x \le 2 \\ 0 & \text{elsewhere} \end{cases}$$

(a) Find $F_X(x)$
(b) A random variable Y is defined es

$$Y = 2X + 1$$

Find $F_Y(y)$
(c) Find $f_Y(y)$.

Solution 4.1 (a) By definition

$$F_x(x) = \int_{-\infty}^x f_x(\tau)d\tau$$

– Case I $x < 0$

$$F_X(x) = \int_{-\infty}^x 0 dt = 0$$

– Case II $0 \le x \le 2$

$$F_X(x) = \int_0^x \frac{1}{2} d\tau = \frac{\tau}{2}\Big|_0^x = \frac{x}{2}$$

– Case III $x > 2$

$$F_x(x) = \int_{-\infty}^x f_x(\tau)d\tau = \int_0^2 \frac{1}{2} d\tau$$

$$= \frac{\tau}{2}\Big|_0^2 = 1$$

Therefore

$$F_X(x) = \begin{cases} 0, & x < 0 \\ \frac{x}{2}, & 0 \le x \le 2 \\ 1, & x > 2 \end{cases}$$

(b)

$$F_Y(y) = P(Y \leq y)$$
$$= P(2X + 1 \leq y)$$
$$= P\left(X \leq \frac{y-1}{2}\right)$$
$$= F_X\left(\frac{y-1}{2}\right)$$

Hence;

$$F_Y(y) = \begin{cases} 0, & \frac{y-1}{2} < 0 \\ \frac{y-1}{4}, & 0 \leq \frac{y-1}{2} \leq 2 \\ 1, & \frac{y-1}{2} > 2 \end{cases}$$

One can easily reorganize the function as

$$F_Y(y) = \begin{cases} 0, & y < 1 \\ \frac{y-1}{4}, & 1 \leqslant y \leqslant 5 \\ 1, & y > 5 \end{cases}$$

(c) $f_Y(y) = \frac{dF_Y(y)}{dy}$

$$f_Y(y) = \begin{cases} 0, & y < 1 \\ \frac{1}{4}, & 1 \leq y \leq 5 \\ 0, & y > 5 \end{cases}.$$

4.5.2 $Y = aX + b$, $a < 0$, *Continuous Random Variables*

Consider X is a random variable. Random variable Y is given as

$$Y = aX + b$$
$$\text{where} \quad a < 0$$

The probability distribution function of the random variable Y;

$$F_Y(y) = P(Y \leq y)$$
$$= P(aX + b \leq y)$$
$$= P\left(X \geq \frac{y-b}{a}\right)$$
$$= 1 - P\left(X \leq \frac{y-b}{a}\right)$$
$$= 1 - F_X\left(\frac{y-b}{a}\right)$$

The probability density function of Y can be derived as,

$$f_Y(y) = \frac{dF_Y(y)}{dy} = \frac{d\left(1 - F_X\left(\frac{y-b}{a}\right)\right)}{dy}$$

$$= \frac{-1}{a} f_X\left(\frac{y-b}{a}\right).$$

Example 4.2 A random variable X has the following probability density function (the same as in Example 4.1)

$$f_Z(x) = \begin{cases} \frac{1}{2}, & 0 \le x \le 2 \\ 0, & \text{elsewhere} \end{cases}$$

Another random variable Y is given as

$$Y = -2X + 1$$

(a) Find $F_y(y)$
(b) Find $f_Y(y)$.

Solution 4.2 (a)
$$F_Y(y) = P(Y \le y)$$
$$= P(-2X + 1 \le y)$$
$$= P\left(X \ge \frac{y-1}{-2}\right)$$
$$= P\left(X \ge \frac{1-y}{2}\right)$$
$$= 1 - P\left(X \le \frac{1-y}{2}\right)$$
$$= 1 - F_X\left(\frac{1-y}{2}\right)$$

Since $F_X(x)$ was already derived in Example 4.1;

$$F_Y(y) = \begin{cases} 1 - 0, & \frac{1-y}{2} < 0 \\ 1 - \left(\frac{1-y}{4}\right), & 0 \le \frac{1-y}{2} \le 2 \\ 1 - 1, & \frac{1-y}{2} > 2 \end{cases}$$

It can be reorganized as

$$F_Y(y) = \begin{cases} 0, & y < -3 \\ \frac{y+3}{4}, & -3 \le y \le 1 \\ 1, & y > 1 \end{cases}$$

(b) By definition

$$f(y) = \frac{dF_Y(y)}{dy}$$

hence,

$$f_Y(y) = \begin{cases} \frac{1}{4}, & -3 \le y \le 1 \\ 0, & \text{elsewhere} \end{cases}.$$

4.5.3 $Y = aX + b$, $a > 0$, *Discrete Random Variables*

Consider that X is a discrete random variable having the probability distribution density function of $f_X(x)$. Another discrete random variable Y is defined

$$Y = aX + b$$

By definition of the discrete probability density function one has

$$\begin{aligned} f_Y(y) &= P(Y = y) \\ &= P(aX + b = y) \\ &= P\left(X = \frac{y - b}{a}\right) \\ &= f_X\left(\frac{y - b}{a}\right). \end{aligned}$$

Example 4.3 Let X be a random variable with the following probability density function

$$f_X(x) = \begin{cases} \frac{1}{4} & \text{if } x = 1 \\ \frac{1}{3} & \text{if } x = 2, 3 \\ \frac{1}{12} & \text{if } x = 4 \end{cases}$$

If $Y = 2X + 3$

(a) Find $F_X(x)$
(b) Find $f_Y(y)$
(c) Find $F_Y(y)$.

Solution 4.3 (a) Since

$$F_X(x_i) = \sum_{k \leq i} f_X(x_k)$$

Therefore;

$$F_X(x) = \begin{cases} 0, & x < 1 \\ \frac{1}{4}, & 1 \leq x < 2 \\ \frac{7}{12}, & 2 \leq x < 3 \\ \frac{11}{12}, & 3 \leq x < 4 \\ 1, & x \geq 4 \end{cases}$$

(b)

$$f_Y(y) = P(Y = y)$$
$$= P(2X + 3 = y)$$
$$= P\left(X = \frac{y-3}{2}\right)$$
$$= f_X\left(\frac{y-3}{2}\right)$$

Hence $f_Y(y)$ can be written as

$$f_Y(y) = \begin{cases} \frac{1}{4}, & \frac{y-3}{2} = 1 \\ \frac{1}{3}, & \frac{y-3}{2} = 2,3 \\ \frac{1}{12}, & \frac{y-3}{2} = 4 \end{cases}$$

After rewriting the function, one has

$$f_Y(y) = \begin{cases} \frac{1}{4}, & y = 5 \\ \frac{1}{3}, & y = 7,9 \\ \frac{1}{12}, & y = 11 \end{cases}$$

(c)

$$F_Y(y) = \begin{cases} 0, & y < 5 \\ \frac{1}{4}, & 5 \le y < 7 \\ \frac{7}{12}, & 7 \le y < 9 \\ \frac{11}{12}, & 9 \le y < 11 \\ 1, & y \ge 11 \end{cases}.$$

4.5.4 $Y = X^2$

Let us assume that the random variable X represents the potential difference across a 1 Ω resistor. Therefore, the dissipated power on the resistor can be expressed as;

$$P = \frac{V^2}{R} = V^2$$

The dissipated power, P, is another random variable, say Y. The probability distribution function of the dissipated power can be written as;

$$\begin{aligned} F_Y(y) &= P(Y \le y) \\ &= P\left(X^2 \le y\right) \\ &= P(-\sqrt{y} \le X \le \sqrt{y}) \\ &= F_X(\sqrt{y}) - F_X(-\sqrt{y}) \end{aligned}$$

The corresponding probability density function is expressed as;

$$\begin{aligned} f_Y(y) &= \frac{d F_Y(y)}{dy} = \frac{1}{2\sqrt{y}} f_X(\sqrt{y}) + \frac{1}{2\sqrt{y}} f_X(-\sqrt{y}) \\ &= \frac{1}{2\sqrt{y}} \left(f_X(\sqrt{y}) + f_X(-\sqrt{y}) \right). \end{aligned}$$

4.5.5 Sum of Two Statistically Independent Continuous Random Variables, $Z = X + Y$

$$\begin{aligned} F_Z(z) &= P(Z \le z) \\ &= P(X + Y \le z) \\ &= \int_{-\infty}^{\infty} f_X(x) \left(\int_{-\infty}^{z-x} f_Y(y)y \right) dx \end{aligned}$$

Since the derivative is a linear operator, if it is employed to both sides of the equation, then the equality still holds.

Therefore

$$f_Z(z) = \int_{-\infty}^{\infty} f_X(x) f_Y(z - x) dx$$

which is a well-known equation in signals and systems area, namely convolution integral.

> Sum of two independent random variables results in another random variable having the probability density function, which equals to the convolution of the individual probability density functions.

4.5.6 Sum of Two Statistically Independent Discrete Random Variables, $Z = X + Y$

The probability density function of Z can be written as

$$f_Z(z) = P(Z = z)$$
$$= P(X + Y = z)$$
$$\text{if} \quad X = k, \text{ then } Y = Z - k$$

since X ad Y are statistically independent random variables then

$$f_Z(z) = \sum_{k=-\infty}^{\infty} P(X = k) P(Y = Z - k)$$

$$= \sum_{k=-\infty}^{\infty} f_X(k) f_Y(Z - k)$$

which is convolution sum equation.

One can easily deduce that a sum of N statistically independent random variables produce a new random variable with a probability density function that can be calculated by means of convolution sum (discrete case) or convolution integral (continuous case);

$$Y = X_1 + X_2 + \cdots + X_n$$
$$f_Y(y) = f_{X_1}(x_1) * f_{X_2}(x_2) * \cdots * f_{X_n}(x_n).$$

Fig. 4.1 Electric circuit
(Example 4.4)

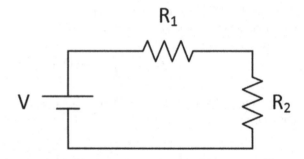

Fig. 4.1 Electric circuit
(Example 4.4)

Example 4.4 Consider the circuit shown in Fig. 4.1. Let X, and Y are random variables representing the value of resistors R_1 and R_2 in $k\Omega$, respectively. X, and Y are uniformly distributed with;

$$f_X(x) = \begin{cases} 5, & 0.9 \le x \le 1.1 \\ 0, & \text{elsewhere} \end{cases}.$$

$$f_Y(y) = \begin{cases} 2.5, & 1.8 \le y \le 2.2 \\ 0, & \text{elsewhere} \end{cases}$$

(a) Derive the probability density function of $R_{eq} = R_1 + R_2$
(b) Calculate $P(3 < R_{eq} \le 3.1)$.

Solution 4.4 (a) Let Z denotes the random variable in order to represent the equivalent resistance, R_{eq}. Since, $R_{eq} = R_1 + R_2$ then $Z = X + Y$, the probability density function of Z can be calculated using convolution integral of the probability density functions of X and Y;

$$\begin{aligned}
f_Z(z) &= f_X(x) * f_Y(y) \\
&= 5\,(u(x - 0.9) - u(x - 1.1)) * 2.5\,(u(y - 1.8) - u(y - 2.2)) \\
&= 12.5((z - 2.7)u(z - 2.7) - (z - 2.9)u(z - 2.9) - (z - 3.1)u(z - 3.1) \\
&\quad + (z - 3.3)u(z - 3.3))
\end{aligned}$$

or, in another form;

$$f_Z(z) = \begin{cases} 12.5(z - 2.7), & 2.7 \le z \le 2.9 \\ 2.5, & 2.9 \le z \le 3.1 \\ -12.5(z - 3.1), & 3.1 \le z \le 3.3 \\ 0, & \text{elsewhere} \end{cases}$$

(b)

$$P(3 < R_{eq} \leq 3.1) = \int_3^{3.1} f_Z(z)dz$$

$$= \int_3^{3.1} 2.5dz$$

$$= 0.25$$

4.5.7 $Z = XY$

$$F_Z(z) = P(Z \leq z) = P(XY \leq z)$$

$$= \int_{-\infty}^{\infty} \int_{-\infty}^{\infty} f_X(x)f_Y(y)u(z - xy)dydx$$

The probability density function can be derived by taking the derivative of both sides with respect to z;

$$f_Z(z) = \int_{-\infty}^{\infty} \int_{-\infty}^{\infty} f_X(x)f_Y(y)\delta(z - xy)dydx$$

$$= \int_{-\infty}^{\infty} f_X(x)f_Y(z/x)\left[\int_{-\infty}^{\infty} \delta(z - xy)dy\right]dx$$

$$= \int_{-\infty}^{\infty} f_X(x)f_Y(z/x)\frac{1}{|x|}dx$$

Since $Z = XY = YX$, therefore the probability density function of Z can be written as;

$$f_Z(z) = \int_{-\infty}^{\infty} f_X(z/y)f_Y(y)\frac{1}{|y|}dy.$$

4.5.8 $Z = \frac{X}{Y}$

The probability distribution function of Z is;

$$
\begin{aligned}
F_Z(z) = P(Z \le z) &= P(X/Y \le z) \\
&= P(X \le yz, y > 0) + P(X \ge yz, y < 0) \\
&= \int_{y=0}^{\infty} \int_{x=-\infty}^{yz} f_X(x) f_Y(y) \mathrm{d}y \mathrm{d}x + \int_{y=-\infty}^{0} \int_{x=yz}^{\infty} f_X(x) f_Y(y) \mathrm{d}y \mathrm{d}x
\end{aligned}
$$

by taking the derivative of both sides with respect to z, the probability density function can be derived as;

$$
\begin{aligned}
f_Z(z) &= \int_0^{\infty} y f_X(yz) f_Y(y) \mathrm{d}y + \int_{-\infty}^0 (-y) f_X(yz) f_Y(y) \mathrm{d}y \\
&= \int_{-\infty}^{\infty} |y| f_X(yz) f_Y(y) \mathrm{d}y.
\end{aligned}
$$

4.5.9   Central Limit Theorem

Let $X_1, X_2, \ldots, X_i, \ldots, X_N$ be statistically independent random variables. If a random variable of Y is defined as;

$$
Y = \sum_{i=1}^{N} X_i
$$

according to central limit theorem, if N is a "large enough", then the probability distribution function of Y approaches a Gaussian distribution.

In practical applications, the central limit theorem is useful to model such a physical system by means of a Gaussian distribution. For instance, a system involving cascade connected subsystems, and the delay time in each subsystem is represented by a random variable X_i. The total delay time can be modelled as a sum of each time delays in each individual subsystems. Obviously, the value of N being "large enough" depends on the dynamics of the system under investigation.

4.6   Problems

4.1 The joint probability density function is given as;

$$f_{XY}(x, y) = Ae^{-5(2x+3y)}u(x)u(y)$$

(a) Find A such that $f_{XY}(x, y)$ is a valid pdf.
(b) Find the marginal probability density functions.
(c) Are X and Y independent random variables?
(d) Compute the probabilities $P\{0 < x \leq 1, 0 < y \leq 2\}$.

4.2 Let X and Y be two continuous random variables with joint probability density function;

$$f_{XY}(x, y) = \begin{cases} cxy(1 + y) & \text{for } 1 \leq x \leq 2 \text{ and } 0 \leq y \leq 4 \\ 0 & \text{elsewhere} \end{cases}$$

(a) Find A such that $f_{XY}(x, y)$ is a valid pdf.
(b) Find the probability $P(0 \leq X \leq 1, -1 \leq Y \leq 3)$
(c) Determine the joint probability distribution function of X and Y, $F_{XY}(x, y)$.
(d) Find the marginal probability density functions, $f_X(x)$ and $f_Y(y)$
(e) Are X and Y independent random variables?

4.3 Consider the following probability density functions of the random variables X and Y;

$$f_X(x) = Ae^{-10x}u(x)$$

$$f_Y(y) = Be^{-4y}u(y)$$

(a) Find A and B such that $f_X(x)$ $f_Y(y)$ are valid probability density functions.
(b) Derive the probability density function of the random variable $Z = X + Y$.
(c) Derive the probability density function of the random variable $Z = X - Y$.
(d) Compute the probability of $P\{0 < x \leq 2, 0 < y \leq 4\}$.

4.4 The joint probability density function of two random variables X and Y is;

$$\begin{aligned} f_{XY}(x, y) = {}& 0.15\delta(x)\delta(y) + 0.05\delta(x - 1)\delta(y) + 0.20\delta(x - 1)\delta(y - 1) \\ & + 0.25\delta(x - 2)\delta(y - 3) + 0.15\delta(x - 3)\delta(y - 5) + 0.10\delta(x - 4)\delta(y - 6) \\ & + 0.10\delta(x - 5)\delta(y - 6) \end{aligned}$$

(a) Derive the marginal probability distribution function of X, i.e. $F_X(x)$.
(b) Derive the marginal probability distribution function of Y, i.e. $F_Y(y)$.
(c) Derive the probability density function of the random variable $Z = X - Y$.
(d) Compute the probability of $P\{0 < x \leq 5, 0 < y \leq 3\}$.

4.5 The joint probability density function of two random variables X and Y is;

$$f_{XY}(x, y) = \begin{cases} Axy^4 & \text{for } 0 \leq x \leq y \leq 1 \\ 0 & \text{elsewhere} \end{cases}$$

(a) Calculate A such that $f_{XY}(x, y)$ is a valid probability density function.
(b) Derive the marginal probability density function of X, i.e. $f_X(x)$.
(c) Derive the marginal probability density function of Y, i.e. $f_Y(y)$.
(d) Are X and Y statistically independent random variables?

4.6 Consider a random variable X having the following probability density function

$$f_X(x) = \begin{cases} \frac{1}{4} & 0 \leq x \leq 4 \\ 0 & \text{elsewhere} \end{cases}$$

(a) Find $F_X(x)$
(b) A random variable Y is defined as

$$Y = 2X + 3$$

Find $F_Y(y)$
(c) Derive $f_Y(y)$.

4.7 Consider a random variable X having the following probability density function

$$f_X(x) = \begin{cases} \frac{1}{3} & 0 \leq x \leq 3 \\ 0 & \text{elsewhere} \end{cases}$$

(a) Find $F_X(x)$
(b) A random variable Y is defined as

$$Y = -2X + 3$$

Find $F_Y(y)$
(c) Derive $f_Y(y)$.

4.8 Let X be a random variable with the following probability density function

$$f_X(x) = \begin{cases} \frac{1}{5}, & x = 0 \\ \frac{1}{4}, & x = 2 \\ \frac{3}{10}, & x = 4 \\ \frac{1}{10}, & x = 6 \\ \frac{3}{20}, & x = 8 \end{cases}$$

If $Y = 2X - 5$

(a) Find $F_X(x)$
(b) Find $f_Y(y)$
(c) Find $F_Y(y)$.

4.9 Let X, Y, Z be a random variables having the following joint probability density function,

$$f_{XYZ}(x, y, z) = Ax(y + z) \quad \text{where} \; -1 \le x, y, z \le 1$$

(a) Find A such that $f_{XYZ}(x, y, z)$ is a proper probability density function
(b) Derive the probability density functions of each pair of variables, i.e. $f_{XY}(x, y)$, $f_{XZ}(x, z)$, and $f_{YZ}(y, z)$
(c) Find the marginal probability density functions of each variable.
(d) Find $F_{XYZ}(x, y, z)$.

4.10 Consider a mixed random variable X with the following probability density function;

$$f_X(x) = \frac{1}{5}\delta(x) + \frac{3}{10}\delta(x - 1) + \frac{2}{10}\delta(x - 2) + \frac{3}{10}\frac{1}{\sqrt{2\pi}}e^{-\frac{x^2}{2}}$$

(a) Derive the corresponding probability distribution function, $F_X(x)$
(b) Calculate the probabilities of $P\{x = 0\}$, $P\{x = 1\}$
(c) Find the probability of $P\{X = 1 | X \ge 1\}$.

Chapter 5
Statistical Analysis of Random Variables

Some reference points are needed to analyze and to make judgments about a given information. For instance, let a boy be 1.78 m of height. Is he tall? To answer this question one needs to know the average height of the set that the boy is included in. If the average height of the set is 1.80 m, one can deduce that the boy is not tall. If the average height is 1.65 m, then the boy could be said tall.

Another example is the following, let a woman was 45 years old when she was dead. If she lived during 19th century, she definitely was old-enough since the expected life time was around 35 years-old at that time. The mean and the variance are two main reference points to gain some understanding about the random variable under investigation. The mean, or the expected value of a random variable is defined as the arithmetic average of the values of the random variable. The variance supplies information how the values of the rando:n variable spread around the mean value. For the equally-likely distributed two random variables with the same mean; the random variable, which las bigger variance, is wider spread around the mean comparing to the other one.

The moment generating functions are powerful tools to calculate mean, variance and some other statistical knowledge about a given random variable.

5.1 Statistical Analysis of One Random Variable

5.1.1 Expected Value and Mean

The expected value, or the statistical average of a random variable is called mean of the corresponding random variable, such that;

$$\mu_X = E[X] = \int_{-\infty}^{\infty} x f_X(x) \mathrm{d}x$$

The expected value of a function of a random variable X is calculated;

$$E[g(X)] = \int_{-\infty}^{\infty} g(x) f_X(x) \mathrm{d}x.$$

Example 5.1 Consider a random variable X with the mean μ_X. Calculate the expected value of the random variable $Y = g(X) = 2X + 1$.

Solution 5.1

$$E[g(X)] = E[2X + 1]$$

$$= \int_{-\infty}^{\infty} (2x + 1) f_X(x) \mathrm{d}x$$

$$= 2 \int_{-\infty}^{\infty} x f_X(x) \mathrm{d}x + \int_{-\infty}^{\infty} f_X(x) \mathrm{d}x$$

$$= 2\mu_X + 1$$

5.1.2 Variance and Standard Deviation

The variance of a random variable supplies information about the variety of the random variable around its expected value. It is formulated as;

$$\mathrm{Var}[X] = \int_{-\infty}^{\infty} (x - E[X])^2 f_X(x) \mathrm{d}x$$

$$= \int_{-\infty}^{\infty} \left(x^2 - 2x E[X] + E[X]^2\right) f_X(x) \mathrm{d}x$$

$$= \int_{-\infty}^{\infty} x^2 f_X(x) \mathrm{d}x - 2E[X] \int_{-\infty}^{\infty} x f_X(x) \mathrm{d}x + E[X]^2 \int_{-\infty}^{\infty} f_X(x) \mathrm{d}x$$

Since,

$$\int_{-\infty}^{\infty} x f_X(x) \mathrm{d}x = E[X]$$

$$\int_{-\infty}^{\infty} f_X(x) \mathrm{d}x = 1$$

then

$$\mathrm{Var}[X] = E\left[X^2\right] - E[X]^2$$

The standard deviation, a non-negative variable notated as σ_X, is the square root of the variance, or it can be stated as

$$\text{Var}[X] = \sigma_X^2$$

5.2 Moment Generating Functions

The nth moment of a random variable X around zero is the expected value of X^n, i.e. $E[X^n]$. Let m_n be denoted as the nth moment of X, then, for the continuous probability density functions

$$m_n = E\left[X^n\right] = \int_{-\infty}^{\infty} x^n f_X(x)\,\mathrm{d}x$$

and for the discrete probability density functions

$$m_n = E\left[X^n\right] = \sum_i x_i^n f_X(x_i)$$

It is obvious that $m_0 = 1$, since the sum off all the possibilities has to be 1. Furthermore, the first moment of X, m_1, equals to the expected value of X. However, in many practical cases, calculations of the higher order moments are not trivial and it is time consuming. Therefore, the following tools are proposed especially to calculate the higher order moments.

5.2.1 Maclaurin Series

A moment generation function $m(t)$ of X is defined as;

$$m(t) = E\left[e^{tx}\right]$$

It is known that the Maclaurin series expression of the exponential function is;

$$e^{tX} = \sum_{k=0}^{\infty} \frac{(tX)^k}{k!} = 1 + tX + \frac{(tX)^2}{2} + \frac{(tX)^3}{3!} + \frac{(tX)^4}{4!} + \cdots$$

The nth moment of a random variable X is calculated by means of the nth order derivative of $m(t)$

$$m_n = \lim_{t \to 0} \frac{d^n m(t)}{dt^n} = \lim_{t-0} \frac{d^n}{dt^n} \left(E\left[e^{tX} \right] \right)$$

Hence, a few first moments are calculated as;

$$m_0 = \lim_{t \to 0} m(t) = \lim_{t \to 0} \left(1 + tX + \frac{(tX)^2}{2!} + \frac{(tX)^3}{3!} + \frac{(tX)^4}{4!} + \cdots \right) = 1$$

$$m_1 = \lim_{t \to 0} \frac{dm(t)}{dt} = \lim_{t \to 0} \left(X + \frac{2X(tX)}{2!} + \frac{3X(tX)^2}{3!} + \frac{4X(tX)^3}{4!} + \cdots \right) = E[X]$$

$$m_2 = \lim_{t \to 0} \frac{d^2 m(t)}{dt^2} = \lim_{t \to 0} \left(\frac{2X^2}{2!} + \frac{6X^2(tX)}{3!} + \frac{12X^2(tX)^2}{4!} + \cdots \right) = E\left[X^2 \right].$$

Example 5.2 Find the expected value and the variance of the following Poisson distribution by employing Maclaurin series technique;

$$f_x(x) = \frac{e^{-\lambda} \lambda^x}{x!} \quad \text{for} \quad x = 0, 1, 2, \ldots$$

Solution 5.2

$$m(t) = E\left[e^{tX} \right]$$

$$= \sum_{x=0}^{\infty} \frac{e^{tx} e^{-\lambda} \lambda^x}{x!}$$

$$= e^{-\lambda} \sum_{x=0}^{\infty} \frac{\left(e^t \lambda \right)^x}{x!}$$

since;

$$\sum_{x=0}^{\infty} \frac{\left(e^t \lambda \right)^x}{x!} = e^{e^t \lambda}$$

$$m(t) = e^{-\lambda} e^{e^t \lambda} = e^{\lambda (e^t - 1)}$$

Thereafter,

$$m_0 = \lim_{t \to 0} m(t) = 1$$

$$m_1 = \lim_{t \to 0} m'(t)$$

$$= \lim_{t \to 0} \left(\lambda e^{-\lambda} e^t e^{e^t \lambda} \right)$$

$$= \lambda$$

$$m_2 = \lim_{t \to 0} m''(t)$$

$$= \lim_{t \to 0} \lambda e^{-\lambda} \left(e^t e^{e^t \lambda} + e^t \lambda e^t e^{e^t \lambda} \right)$$

$$= \lambda^2 + \lambda.$$

5.2.2   Characteristic Function

The characteristic function of a random variable X is given as:

$$\Phi_X(\omega) = E\left[e^{j w x} \right]$$

$$= \int_{-\infty}^{\infty} e^{j w x} f_X(x) \mathrm{d}x.$$

which is exactly the Fourier transform of the probability density function $f_X(x)$ with the negative sign of ω. Dirichlet conditions, the sufficient but not necessary conditions for which a continuous function having its Fourier transform are:

(i) $f_X(x)$ has to be bounded with finite numbers of maximum and minimum as its lower and upper limits.

$$K_1 \leq f_X(x) \leq K_2 \quad \text{for all } x$$

where K_1 and K_2 are finite numbers.

(ii) $f_X(x)$ has to be a stable function, that is

$$\int_{-\infty}^{\infty} |f_X(x)| \, \mathrm{d}x < \infty$$

The main advantage of the characteristic function regarding to find moments is that the characteristic function of a random variable defined by a continuous probability density function and its derivatives exist.

The nth moment of a probability density function is calculated as;

$$m_n = \lim_{w \to 0} (-j)^n \frac{d^n \Phi_X(\omega)}{dw^n}.$$

Example 5.3 Considering the following exponential probability density function of a random variable X;

$$f_X(x) = \lambda e^{-\lambda x} u(x), \quad \text{where } \lambda > 0$$

(a) Derive the characteristic function of the related distribution
(b) Calculate the first moment, m_1
(c) Calculate the second moment, m_2.

Solution 5.3 (a)

$$\Phi_X(\omega) = E\left[e^{jwx}\right] = \int_{-\infty}^{\infty} e^{jwx} f_X(x) dx$$

$$= \int_{-\infty}^{\infty} e^{(jwx)} \lambda e^{-\lambda x} u(x) dx$$

$$= \int_{0}^{\infty} e^{(jwx)} \lambda e^{-\lambda x} dx$$

$$= \int_{0}^{\infty} \lambda e^{(jw-\lambda)x} dx$$

$$= \frac{\lambda}{jw - \lambda} e^{(jw-\lambda)x} \big|_0^{\infty}$$

$$= \frac{\lambda}{\lambda - jw}$$

(b)

$$m_1 = \lim_{w \to 0} (-j)^1 \frac{d\Phi_X(\omega)}{dw}$$

$$= \lim_{w \to 0} (-j) \frac{j\lambda}{(\lambda - jw)^2}$$

$$= \lim_{w \to 0} -j^2 \frac{\lambda}{\lambda^2}$$

$$= \frac{1}{\lambda}$$

(c)

$$m_2 = \lim_{w \to 0} (-j)^2 \frac{d^2 \Phi_X(\omega)}{dw^2}$$

$$= \lim_{w \to 0} (-j)^2 \frac{j^2 \lambda (2(\lambda - jw))}{(\lambda - jw)^4}$$

$$= \frac{\lambda (2\lambda)}{\lambda^4}$$

$$= \frac{2}{\lambda^2}.$$

5.3 Statistical Analysis of Multiple Random Variables

5.3.1 Normalized Joint Moments

The normalized joint moment of a multivariate joint probability distribution can be expressed as;

$$m_{n,k} = E[X^n Y^k] = \int_{-\infty}^{\infty} \int_{-\infty}^{\infty} x^n y^k f_{XY}(x, y) dxdy$$

Considering two random variables, namely X and Y. The summation of n and k, i.e. $n + k$, is called *order of the moments*. Hence, the first order normalized joint moments are;

- $m_{1,0} = E[XY^0] = E[X]$ which is the expected value of the random variable X.
- $m_{0,1} = E[X^0Y] = E[Y]$ where $E[Y]$ is the expected value of the random variable Y.

The second order normalized joint moments are;

- $m_{2,0} = E[X^2 Y^0] = E[X^2]$
- $m_{0,2} = E[X^0 Y^2] = E[Y^2]$
- $m_{1,1} = E[XY]$

 $m_{1,1}$ is defined as the correlation between X and Y, which is notated as R_{XY}.

$$R_{XY} = E[XY] = \int_{-\infty}^{\infty} \int_{-\infty}^{\infty} xy f_{XY}(x, y) dxdy.$$

Properties of the Correlation Function, R_{XY}

- If X and Y are uncorrelated random variables, then
  $R_{XY} = E[XY] = E[X]E[Y]$
- If X and Y are statistically independent random variables, then the joint probability density function can be written as;

$$f_{XY}(x, y) = f_X(x)f_Y(y)$$

$$\begin{aligned}
R_{XY} = E[XY] &= \int_{-\infty}^{\infty} \int_{-\infty}^{\infty} xy f_{XY}(x, y) dx dy \\
&= \int_{-\infty}^{\infty} \int_{-\infty}^{\infty} x f_X(x) y f_Y(y) dx dy \\
&= \int_{-\infty}^{\infty} x f_X(x) dx \int_{-\infty}^{\infty} y f_Y(y) dy \\
&= E[X]E[Y]
\end{aligned}$$

Therefore;

> If X and Y are statistically independent random variables, then they are uncorrelated. On the other hand, apart from Gaussian random variables, if X and Y are uncorrelated random variables, one cannot state that they are statistically independent.

- X and Y random variables are called *orthogonal* if the correlation function of these two random variables is zero, i.e. $R_{XY} = 0$.

Example 5.4 Considering the following joint probability density function of the random variables X, and Y;

$$f_{XY}(x, y) = \begin{cases} kxy, & 0 < x < 4; 0 < y < 1 \\ 0, & \text{Elsewhere} \end{cases}$$

(a) Find k in order to satisfy a proper probability density function
(b) Derive the marginal probability density functions, $f_X(x)$, and $f_Y(y)$
(c) Calculate $E[X]$, and $E[Y]$
(d) Find the correlation R_{XY}
(e) Are X and Y correlated?
(f) Are X and Y statistically independent?

Solution 5.4 (a)

$$\int_{-\infty}^{\infty} \int_{-\infty}^{\infty} f_{XY}(x, y) dx dy = 1$$

$$\int_{x=0}^{4} \int_{y=0}^{1} kxy dx dy = 1$$

$$k \left(\frac{x^2}{2} \right) \Big|_0^4 \left(\frac{y^2}{2} \right) \Big|_0^1 = 1$$

$$k \left(\frac{16}{2} \right) \left(\frac{1}{2} \right) = 1$$

$$\text{then} \quad k = \frac{1}{4}$$

(b)

$$f_X(x) = \int_{-\infty}^{\infty} f_{XY}(x, y) dy$$

$$= \int_0^1 \frac{1}{4} xy dy$$

$$= \frac{1}{4} x \left(\frac{y^2}{2} \right) \Big|_0^1$$

$$= \frac{x}{8}$$

$$f_Y(y) = \int_{-\infty}^{\infty} f_{XY}(x, y) dx$$

$$= \int_0^4 \frac{1}{4} xy dx$$

$$= \frac{1}{4} y \left(\frac{x^2}{2} \right) \Big|_0^4$$

$$= 2y$$

(c)

$$E[X] = \int_{-\infty}^{\infty} x f_X(x) dx$$

$$= \int_0^4 x \frac{x}{8} dx$$

$$= \left(\frac{x^3}{24} \right) \Big|_0^4$$

$$= \frac{8}{3}$$

$$E[Y] = \int_{-\infty}^{\infty} y f_Y(y) dy$$

$$= \int_0^1 2y^2 dy$$

$$= \left(\frac{2y^3}{3}\right)\Big|_0^1$$

$$= \frac{2}{3}$$

(d)

$$R_{XY} = E[XY]$$

$$= \int_{-\infty}^{\infty} \int_{-\infty}^{\infty} xy f_{XY}(x, y) dx dy$$

$$= \int_{x=0}^4 \int_{y=0}^1 xy \frac{1}{4} xy \, dx dy$$

$$= \frac{1}{4} \left(\frac{x^3}{3}\right)\Big|_0^4 \left(\frac{y^3}{3}\right)\Big|_0^1$$

$$= \frac{16}{9}$$

(e) If X and Y are uncorrelated, then;

$$R_{XY} = E[XY] = E[X]E[Y]$$

$$\frac{16}{9} = \frac{8}{3}\frac{2}{3}$$

then, yes X and Y are uncorrelated random variables.

(f) If X and Y are statistically independent, then;

$$f_{XY}(x, y) = f_X(x) f_Y(y)$$

$$\frac{1}{4} xy = \frac{x}{8} 2y$$

therefore, X and Y are statistically independent random variables.

5.3.2 Joint Moments Around the Expected Values

The joint moments around the expected values are called joint central moments. Considering two random variables, namely X and Y, the joint central moment can be calculated as;

$$\mu_{n,k} = E[(X - \mu_X)^n (Y - \mu_Y)^k]$$
$$= \int_{-\infty}^{\infty} \int_{-\infty}^{\infty} (x - \mu_X)^n (y - \mu_Y)^k f_{XY}(x, y) dx dy$$

The summation of n and k, i.e. $n + k$, is called *order of the moments*. Thereafter, the first order joint moments are;

- $\mu_{1,0} = E[X - \mu_X]$

$$\mu_{1,0} = E[(X - \mu_X)(Y - \mu_Y)^0]$$
$$= \int_{-\infty}^{\infty} (x - \mu_X) f_X(x) dx$$
$$= \int_{-\infty}^{\infty} x f_X(x) dx - \int_{-\infty}^{\infty} \mu_X f_X(x) dx$$
$$= \mu_X - \mu_X$$
$$= 0$$

- $\mu_{0,1} = E[Y - \mu_Y]$

$$\mu_{0,1} = E[(X - \mu_X)^0 (Y - \mu_Y)]$$
$$= \int_{-\infty}^{\infty} (y - \mu_Y) f_Y(y) dy$$
$$= \int_{-\infty}^{\infty} y f_Y(y) dy - \int_{-\infty}^{\infty} \mu_Y f_Y(y) dy$$
$$= \mu_Y - \mu_Y$$
$$= 0$$

The second order normalized joint moments are;

- $\mu_{2,0} = E[(X - \mu_X)^2]$

$$\mu_{2,0} = E[(X - mu_X)^2(Y - \mu_Y)^0]$$

$$= \int_{-\infty}^{\infty} (x - \mu_X)^2 f_X(x)dx$$

$$= \int_{-\infty}^{\infty} x^2 f_X(x)dx - \int_{-\infty}^{\infty} 2x\mu_X f_X(x)dx + \int_{-\infty}^{\infty} \mu_X^2 f_X(x)dx$$

$$= E[X^2] - 2\mu_X^2 + \mu_X^2$$

$$= E[X^2] - \mu_X^2$$

$$= Var[X]$$

$$= \sigma_X^2$$

which is the variance of the random variable X

- $\mu_{0,2} = E[(Y - \mu_Y)^2]$

$$\mu_{0,2} = E[(X - mu_X)^0(Y - \mu_Y)^2]$$

$$= \int_{-\infty}^{\infty} (y - \mu_Y)^2 f_Y(y)dy$$

$$= \int_{-\infty}^{\infty} y^2 f_Y(y)dy - \int_{-\infty}^{\infty} 2y\mu_Y f_Y(y)dy + \int_{-\infty}^{\infty} \mu_Y^2 f_Y(y)dy$$

$$= E[Y^2] - 2\mu_Y^2 + \mu_Y^2$$

$$= E[Y^2] - \mu_Y^2$$

$$= Var[Y]$$

$$= \sigma_Y^2$$

which is the variance of the random variable Y

- $\mu_{1,1} = E[(X - \mu_X)(Y - \mu_Y)]$

$$\mu_{1,1} = E[(X - \mu_X)(Y - \mu_Y)]$$

$$= \int_{-\infty}^{\infty} \int_{-\infty}^{\infty} (x - \mu_X)(y - \mu_Y) f_{XY}(x, y) dx dy$$

$$= \int_{-\infty}^{\infty} \int_{-\infty}^{\infty} xy f_{XY}(x, y) dx dy - \int_{-\infty}^{\infty} \int_{-\infty}^{\infty} \mu_X y f_{XY}(x, y) dx dy$$

$$- \int_{-\infty}^{\infty} \int_{-\infty}^{\infty} x \mu_Y f_{XY}(x, y) dx dy + \int_{-\infty}^{\infty} \int_{-\infty}^{\infty} \mu_X \mu_Y f_{XY}(x, y) dx dy$$

$$= R_{XY} - \mu_X \mu_Y - \mu_X \mu_Y + \mu_X \mu_Y$$

$$= R_{XY} - \mu_X \mu_Y$$

The second order joint moment, $\mu_{1,1}$ is called *covariance* of the random variables X and Y, and it is notated as C_{XY}. Hence,

$$C_{XY} = \mu_{1,1}$$
$$= R_{XY} - E[X]E[Y].$$

Properties of the Covariance Function, C_{XY}

- If X and Y are statistically independent or uncorrelated random variables, then

$$R_{XY} = E[XY] = E[X]E[Y]$$
$$C_{XY} = R_{XY} - E[X]E[Y]$$
$$C_{XY} = 0$$

- If X and Y are orthogonal random variables,

$$R_{XY} = 0$$
$$C_{XY} = R_{XY} - E[X]E[Y]$$
$$C_{XY} = -E[X]E[Y].$$

Correlation Coefficient, ρ_{XY}

The normalized 2nd order central moment can be calculated as;

$$\rho_{XY} = \frac{\mu_{1,1}}{\sqrt{\mu_{2,0}\mu_{0,2}}} = \frac{C_{XY}}{\sigma_X \sigma_Y}$$

where ρ_{XY} is known as the correlation coefficient of the random variables X and Y. The range of the correlation coefficient is

$$-1 \leq \rho_{XY} \leq 1$$

If $|\rho_{XY}|$ approaches zero, then it can be stated that X and Y are less correlated random variables, if $|\rho_{XY}|$ goes one it is said that X and Y are highly correlated random variables.

Example 5.5 Consider the following joint probability density function of the random variables X, and Y;

$$f_{XY}(x, y) = e^{-(x+y)}u(x)u(y)$$

(a) Derive the marginal probability density functions, $f_X(x)$, and $f_Y(y)$
(b) Calculate $E[X]$, and $E[Y]$
(c) Calculate $Var[X]$, and $Var[Y]$
(d) Find the correlation R_{XY}
(e) Find the covariance C_{XY}
(f) Calculate the correlation coefficient ρ_{XY}
(g) Are X and Y correlated?
(h) Are X and Y statistically independent?

Solution 5.5 (a)

$$f_X(x) = \int_{-\infty}^{\infty} f_{XY}(x, y)dy$$

$$= \int_{0}^{\infty} e^{-(x+y)}dy$$

$$= e^{-(x+y)}\Big|_{0}^{\infty}$$

$$= e^{-x}u(x)$$

$$f_Y(y) = \int_{-\infty}^{\infty} f_{XY}(x, y)dx$$

$$= \int_{0}^{\infty} e^{-(x+y)}dx$$

$$= e^{-(x+y)}\Big|_{0}^{\infty}$$

$$= e^{-y}u(y)$$

(b)

$$E[X] = \int_{-\infty}^{\infty} x f_X(x) dx$$
$$= \int_0^{\infty} x e^{-x} dx$$
$$= -e^{-x}(x+1)\big|_0^{\infty}$$
$$= 1$$

$$E[Y] = \int_{-\infty}^{\infty} y f_Y(y) dy$$
$$= \int_0^{\infty} y e^{-y} dx$$
$$= -e^{-y}(y+1)\big|_0^{\infty}$$
$$= 1$$

(c)

$$Var[X] = \int_{-\infty}^{\infty} (x - \mu_X)^2 f_X(x) dx$$
$$= \int_0^{\infty} (x-1)^2 e^{-x} dx$$
$$= -e^{-x}(x^2+1)\big|_0^{\infty}$$
$$= 1$$

$$Var[Y] = \int_{-\infty}^{\infty} (y - \mu_Y)^2 f_Y(y) dy$$
$$= \int_0^{\infty} (y-1)^2 e^{-y} dy$$
$$= -e^{-y}(y^2+1)\big|_0^{\infty}$$
$$= 1$$

(d)

$$R_{XY} = E[XY]$$
$$= \int_{-\infty}^{\infty} \int_{-\infty}^{\infty} xy f_{XY}(x, y) dx dy$$
$$= \int_{x=0}^{\infty} \int_{y=0}^{\infty} xy e^{-(x+y)} dx dy$$
$$= (x+1)(y+1) e^{-(x+y)}\big|_0^{\infty} \big|_0^{\infty}$$
$$= 1$$

(e)
$$C_{XY} = R_{XY} - E[X]E[Y]$$
$$C_{XY} = 1 - 1$$
$$= 0$$

(f)
$$\rho_{XY} = \frac{\mu_{1,1}}{\sqrt{\mu_{2,0}\mu_{0,2}}} = \frac{C_{XY}}{\sigma_X \sigma_Y}$$
$$= 0$$

(g)   If X and Y are uncorrelated, then;

$$R_{XY} = E[XY] = E[X]E[Y]$$
from the previous part
$$1 = 1 \times 1$$

hence, X and Y are uncorrelated random variables.

(h)   If X and Y are statistically independent, then;

$$f_{XY}(x, y) = f_X(x)f_Y(y)$$
from the previous part
$$e^{-(x+y)}u(x)u(y) = e^{-(x)}u(x)e^{-(y)}u(y)$$

then, yes X and Y are statistically independent random variables.

5.3.3 Expected Operations of Functions of Random Variables

Consider that X and Y are two random variables;

- The expected value of $g(X)$ can be given as;

$$E[g(X)] = \sum_{\text{for all } i} g(x_i)f_X(x_i) \quad \text{Discrete case}$$

$$E[g(X)] = \int_{-\infty}^{\infty} g(x)f_X(x)dx \quad \text{Continues case}$$

- The expected value of $g(X, Y)$ can be given as;

$$E[g(X, Y)] = \sum_{\text{for all } i} \sum_{\text{for all } j} g(x_i, y_j)f_{X,Y}(x_i, y_j) \quad \text{Discrete case}$$

$$E[g(X, Y)] = \int_{-\infty}^{\infty}\int_{-\infty}^{\infty} g(x, y)f_{XY}(x, y)dxdy \quad \text{Continues case}$$

Some Important Properties of Expected Value and Variance Operations

- If $g(X) \geq h(X)$ for all x, then;

$$E[g(X)] \geq E[h(X)]$$

-

$$E[aX + bY + c] = aE[X] + bE[y] + c$$

-

$$Var[aX + b] = a^2 Var[X]$$

-

$$Var[aX + bY] = a^2 Var[X] + b^2 Var[Y] + 2abC_{XY}$$

-

$$C_{XY} \leq \sqrt{Var[X]Var[Y]}.$$

5.4 Problems

5.1 The joint probability density function is given as;

$$f_{XY}(x, y) = \frac{1}{4} e^{-\frac{1}{2}(x+y)} u(x)u(y)$$

(a) Derive the marginal probability density functions, $f_X(x)$, and $f_Y(y)$
(b) Calculate $E[X]$, and $E[Y]$
(c) Calculate $Var[X]$, and $Var[Y]$
(d) Find the correlation R_{XY}
(e) Find the covariance C_{XY}
(f) Calculate the correlation coefficient ρ_{XY}
(g) Are X and Y correlated?
(h) Are X and Y statistically independent?

5.2 Considering the following joint probability density function of the random variables X, and Y;

$$f_{XY}(x, y) = \begin{cases} \frac{1}{8}, & -2 < x < 2; 0 < y < 2 \\ 0, & \text{Elsewhere} \end{cases}$$

(a) Calculate the expected value of $g(X, Y) = X + Y$, i.e. $E[g(X, Y)]$.
(b) Calculate the expected value of $g(X, Y) = X^2 + Y$.

5.3 Considering the given joint probability density function,

$$f_{XY}(x, y) = 4e^{-2(x+y)}u(x)u(y)$$

(a) Calculate the expected value of $g(X, Y) = e^{-(X+Y)}$
(b) Calculate the variance of $g(X, Y) = e^{-(X+Y)}$

5.4 Considering the following statistically independent random variables;

Random variable	Mean	Variance
X	2	4
Y	1	9
Z	-4	16

Calculate the mean values and variances of the following functions;

(a) $g(X, Y) = 2X - 3Y$
(b) $g(Y, Z) = Y + 4Z$
(c) $g(X, Z) = 3X - Z$
(d) $g(X, Y, Z) = X + Y + Z$ [Hint: assign a new random variable such that $W = X + Y$, then consider $g(W, Z)$].

5.5 Let X, Y, Z be a random variables having the following joint probability density function,

$$f_{XYZ}(x, y, z) = Kx(y + z) \quad \text{where} \ -2 \le x, y, z \le 2$$

(a) Find K such that $f_{XYZ}(x, y, z)$ is a proper probability density function
(b) Derive the probability density functions of each pair of variables, i.e. $f_{XY}(x, y)$, $f_{XZ}(x, z)$, and $f_{YZ}(y, z)$,
(c) Find the marginal probability density functions of each variable.
(d) Discuss whether X, Y, Z are independent random variables or not.

5.6 Consider a mixed random variable X with the following probability density function;

$$f_X(x) = \frac{2}{5}\delta(x - 1) + \frac{1}{10}\delta(x) + \frac{3}{10}\delta(x - 1) + \frac{1}{5}\frac{1}{\sqrt{2\pi}}e^{-\frac{x^2}{2}}$$

(a) Derive the corresponding probability distribution function, $F_X(x)$
(b) Calculate the probabilities of $P\{x = -1\}$, $P\{x = 1\}$
(c) Find the probability of $P\{X = 0 | X \le 1\}$
(d) Calculate $E[X]$ and $Var[X]$.

5.7 Assume that X and Y have the probability density function of

$$f_{XY}(x, y) = Axy^2 \quad \text{where} \ -1 \le x, y \le 2$$

(a) Find A such that $f_{XY}(x, y)$ is a proper probability density function

(b) Calculate the probability density function of X
(c) Calculate the probability density function of Y
(d) Discuss whether X, Y are independent random variables or not.

5.8 Let X and Y be random variables with the probability density function

$$f_{XY}(x, y) = Axy^2 \quad \text{where} \; -1 \leq x \leq y \leq 2$$

(a) Find A such that $f_{XY}(x, y)$ is a proper probability density function
(b) Calculate the probability density function of X
(c) Calculate the probability density function of Y
(d) Discuss whether X, Y are independent random variables or not.

Part II
Random Processes

Chapter 6
Random Processes

In a general point of view, a system might have a random nature itself; or it could be deterministic but including an undesired random noise part; or such a process consists of both random main part and random noisy part. In all these cases, the system has to be analyzed taking into consideration its randomness and the time phenomena.

6.1 Random Processes

Let us recall that a random variable X is a function mapping all the possible outcomes s into the $\Re$ space as introduced in Chap. 2. Now, we will move one step forward and define a new function depending on the possible outcomes, s, and the time, t. Such a function denoted as $X(s, t)$ is called a random process.

The following notations can be proposed considering a random process;

- $X(s, t)$ is a family of functions, which is called *ensemble.*
- If s is fixed at a certain value of s_1, then $X(s_1, t)$ is a deterministic function of t, which is called *realization.*
- If t is hold at a certain value of t_1, then $X(s, t_1)$ is just a random variable frozen at t_1, which is called *time sample.*
- If both s and t are fixed at certain values, then $X(s, t)$ is just a number.

© The Author(s), under exclusive license to Springer Nature Switzerland AG 2022
M. Catak et al., *Probability and Random Variables for Electrical Engineering*,
Studies in Systems, Decision and Control 390,
https://doi.org/10.1007/978-3-030-82922-3_6

6.1.1   Probability Functions Associated with a Random Process

The first order probability distribution function of a random process $X(s, t)$ at a specific tine t_1 is explained as;

$$F_X(x_1; t_1) = P(X(t_1) \leq x_1)$$

The second order joint probability distribution function is defined;

$$F_X(x_1, x_2; t_1, t_2) = P(X(t_1) \leq x_1, X(t_2) \leq x_2)$$

Finally, the general form of the joint probability distribution function for the number of N random variables can be written as;

$$F_X(x_1, x_2, \ldots, x_N; t_1, t_2, \ldots, t_N) = P(X(t_1) \leq x_1, X(t_2) \leq x_2, \ldots, X(t_N) = x_N)$$

Thereafter, the corresponding joint probability density functions are obtained using proper differentiate of the joint probability distribution functions.

$$f_X(x_1; t_1) = \frac{d F_X(x_1; t_1)}{dx_1}$$

$$f_X(x_1, x_2; t_1, t_2) = \frac{\partial^2 F_X(x_1, x_2; r_1, t_2)}{\partial x_1 \partial x_2}$$

$$f_X(x_1, x_2, \ldots, x_N; t_1, t_2, \ldots, t_N) = \frac{\partial^N F_X(x_1, x_2, \ldots, x_N; t_1, t_2, \ldots, t_N)}{\partial x_1 \partial x_2 \ldots \partial x_N}$$

6.1.2   Classification of Random Processes

Discrete versus Continuous Random Processes
A random process $X(s, t)$ is classified into four subgroups according to its behavior with respect to s, and t. The possible combinations of discrete and continuous cases of s and t are summarized in Table 6.1.

Table 6.1 Classification of a random process according to s and t

	Discrete in samples	Continuous in samples
Discrete in time	Discrete random sequence	Continuous random sequence
Continuous in time	Discrete random process	Continuous random process

In this chapter, random processes, where the time t parameter is continuous, are discussed.

Deterministic versus Stochastic Random Processes

Consider a random variable $X(s, t)$ for which s is fixed at a certain value. Therefore it can be written as $X(t) = X(s, t)$. If $X(t)$ is determined, that is the values of $X(t)$ can be estimated for any value of t, then such a random process is called deterministic random process. Otherwise, the random process is nondeterministic, or it can be said as a stochastic process.

Statistically Independent Random Processes

Random processes are called statistically independent if their joint probability density function can be obtained by multiplications of their individual joint probability density functions in any order $N < \infty$. Let $X(t)$ and $Y(t)$ be two random processes. If they are statistically independent, then it is written that;

$$f_{XY}(x_1, x_2, \ldots, x_N, y_1, y_2, \ldots, y_M; t_1, t_2, \ldots, t_N, \tau_1, \tau_2, \ldots, \tau_M)$$

$$= f_X(x_1, x_2, \ldots, x_N; t_1, t_2, \ldots, t_N) \, f_Y(y_1, y_2, \ldots, y_M; \tau_1, \tau_2, \ldots, \tau_M)$$

Stationary versus Non-stationary Random Processes

First Order Stationary Random Process

A random process $X(s, t)$ is called first order stationary if its first order probability density function holds the same function with respect to any shift on the time axis. That is;

$$f_X(x_1; t_1) = f_X(x_1; t_1 + \Delta t)$$

Therefore, the expected value of a such process is constant with respect to time,

$$E[X(t)] = \mu_X = \text{constant}$$

Having a constant expected value is the trade mark of a first order stationary random process.

Wide-sense Stationary (WSS) Random Process

A random process is called second order stationary if its second order probability density function remains unchanged with respect to any shift on the time axis;

$$f_X(x_1, x_2; t_1, t_2) = f_X(x_1, x_2; t_1 + \Delta t, t_2 + \Delta t)$$

The autocorrelation function of a random process is defined as;

$$R_{XX}(t_1, t_2) = E[X(t_1) X(t_2)]$$

If a new variable τ is introduced as;

$$\tau = t_2 - t_1$$

then;

$$R_{XX}(t_1, t_2) = R_{XX}(t_1, t_1 + \tau) = R_{XX}(\tau)$$

This means that if a random process is second order stationary, then its autocorrelation function depends merely on τ.

In a more general way, a random process is called *wide-sense stationary* if the following conditions hold;

$$E[X(t)] = \mu_X = \text{constant}$$
$$E[X(t)X(t + \tau)] = R_{XX}(\tau)$$

Ergodic versus Non-Ergodic Random Processes

The time average of a function $f(t)$ is calculated as;

$$A[f(t)] = \lim_{t \to \infty} \frac{1}{2T} \int_{-T}^{T} f(t)\mathrm{d}t$$

Then, two important functions of a random process are introduced;

- The time average of the realization of a random process

$$\mu_{X,t} = A[x(t)] = \lim_{t \to \infty} \frac{1}{2T} \int_{-T}^{T} x(t)\mathrm{d}t$$

- The time autocorrelation function of a random process

$$R_{XX,t} = A[x(t)x(t + \tau)] = \lim_{t \to \infty} \frac{1}{2T} \int_{-T}^{T} x(t)x(t + \tau)\mathrm{d}t$$

Finally, a random process is called *ergodic* if the following conditions are valid;

(i) The time average of a random process equals to its statistical average (i.e. its expected value)

$$\mu_{X,t} = \mu_X$$

(ii) The time autocorrelation function is equal to the statistical autocorrelation function

$$R_{XX}(\tau) = R_{XX,t}(\tau)$$

Example 6.1 Consider the following random process;

$$X(t) = K \cos(wt + \phi)$$

where K and w are constants; ϕ is a uniformly distributed random variable on $-\pi < \phi < \pi$

(a) Find the mean value of the random process, $E[X(t)]$
(b) Derive the autocorrelation function, $R_{XX}(t, t + \tau)$
(c) Is $X(t)$ wide-sense stationary (WSS)

Solution 6.1 (a) Since, $f_\Phi(\phi) = \frac{1}{2\pi}$ where $-\pi < \phi < \pi$

$$\begin{aligned}
E[X(t)] &= \int_{-\pi}^{\pi} K \cos(wt + \phi) \frac{1}{2\pi} d\phi \\
&= -\frac{K}{2\pi} \sin(wt + \phi) \mid_{-pi}^{\pi} \\
&= -\frac{K}{2\pi} (\sin(wt + \pi) - \sin(wt - \pi)) \\
&= -\frac{K}{2\pi} (-\sin(wt) + \sin(wt)) \\
&= 0 \quad \text{which is a constant.}
\end{aligned}$$

(b)
$$R_{XX}(t, t + \tau) = E[K \cos(wt + \phi) K \cos(wt + w\tau + \phi)]$$
$$\text{using } \cos(A) \cos(B) = \frac{1}{2}(\cos(A + B) + \cos(A - B))$$
$$= \frac{K^2}{2} E[\cos(w\tau)] + \frac{K^2}{2} E[\cos(2wt + w\tau + 2\phi)]$$
by following the integral operation discussed in part (a)
$$E[\cos(2wt + w\tau + 2\phi)] = 0; \quad \text{therefore}$$
$$R_{XX}(\tau) = \frac{K^2}{2} \cos(w\tau)$$

(c) Since mean value of the random process $X(t)$ is constant, which is calculated as 0; and its autocorrelation function depends merely on τ, $X(t)$ is a wide-sense stationary (WSS) random process.

6.2 Correlation Functions

6.2.1 Autocorrelation Function, $R_{XX}(t_1, t_2)$

The autocorrelation of a random process can be simply defined as;

$$R_{XX}(t_1, t_2) = E[X(t_1)X(t_2)]$$

let define a variable τ be $\tau = t_2 - t_1$, then

$$R_{XX}(t, t + \tau) = E[X(t)X(t + \tau)]$$

If the random process is a wide-sense stationary (WSS) random process, then the autocorrelation function is depends only the parameter of τ.

Properties of $R_{XX}(\tau)$ of a WSS random process
(i)
$$R_{XX}(0) = E[X^2(t)]$$

(ii) The maximum value of the autocorrelation function is at $\tau = 0$.

$$|R_{XX}(\tau)| \leq R_{XX}(0)$$

(iii) The autocorrelation function is an even-symmetric function.

$$R_{XX}(-\tau) = R_{XX}(\tau)$$

(iv) If $X(t)$ is an ergodic process with no periodic components, and having a nonzero expected value, then

$$\lim_{\tau \to \infty} R_{XX}(\tau) = (E[X(t)])^2 = \mu_X^2$$

6.2.2 Cross-Correlation Function, $R_{XY}(t_1, t_2)$

The cross-correlation function of random variables $X(t)$, and $Y(t)$ can be written as;

$$R_{XY}(t_1, t_2) = E[X(t_1)Y(t_2)]$$

let $\tau = t_2 - t_1$, then

$$R_{XY}(t, t + \tau) = E[X(t)Y(t + \tau)]$$

If $X(t)$ and $Y(t)$ are jointly wide-sense stationary random process, the cross-correlation function depends only on τ;

$$R_{XX}(\tau) = E[X(t)Y(t + \tau)]$$

(i) If $X(t)$ and $Y(t)$ are orthogonal random processes

$$R_{XY}(t, t + \tau) = 0$$

(ii) If $X(t)$ and $Y(t)$ are statistically independent random processes

$$R_{XY}(t, t + \tau) = E[X(t)]E[Y(t + \tau)]$$

Properties of $R_{XY}(\tau)$ of a WSS random process
(i)
$$R_{XY}(\tau) = \mu_X \mu_Y$$

which is a constant
(ii) The cross-correlation function is a bounded function as;

$$|R_{XY}(\tau)| \leq \sqrt{R_{XX}(0)R_{YY}(0)}$$

(iii) The cross-correlation function is a symmetric function.

$$R_{XY}(-\tau) = R_{YX}(\tau)$$

(iv) If $X(t)$ is a WSS random process and it can be differentiable wwith respect to time. Let a random process $Y(t) = \frac{d(X(t))}{dt}$;

$$R_{XY}(\tau) = \frac{d}{dt}R_{XX}(\tau)$$
$$R_{YY}(\tau) = \frac{d^2}{dt^2}R_{XX}(\tau)$$

6.3 Covariance Functions

6.3.1 Autocovariance Function, $C_{XX}(t_1, t_2)$

The autocovariance of a random process can be simply defined as;

$$C_{XX}(t_1, t_2) = E[(X(t_1) - E[X(t_1)])\,(X(t_2) - E[X(t_2)])]$$

Let us define a variable τ be $\tau = t_2 - t_1$. Then

$$
\begin{aligned}
C_{XX}(t, t+\tau) &= E[(X(t) - E[X(t)])\,(X(t+\tau) - E[X(t+\tau)])] \\
&= E[X(t)X(t+\tau) - E[X(t)]X(t+\tau) - X(t)E[X(t+\tau)] + E[X(t)]E[X(t+\tau)]]
\end{aligned}
$$

Since expectation is a linear operator, one has

$$
\begin{aligned}
C_{XX}(t, t+\tau) &= E[X(t)X(t+\tau)] - E[X(t)]E[X(t+\tau)] \\
&\quad - E[X(t)]E[X(t+\tau)] + E[X(t)]E[X(t+\tau)] \\
&= E[X(t)X(t+\tau)] - E[X(t)]E[X(t+\tau)] \\
&= R_{XX}(t, t+\tau) - E[X(t)]E[X(t+\tau)]
\end{aligned}
$$

6.3.2 Cross-covariance Function, $C_{XX}(t_1, t_2)$

Let $X(t)$ and $Y(t)$ be two random process. The cross-covariance function for $X(t)$ and $Y(t)$ is derived as;

$$C_{XX}(t_1, t_2) = E[(X(t_1) - E[X(t_1)])\,(X(t_2) - E[X(t_2)])]$$

let define a variable τ be $\tau = t_2 - t_1$, then

$$
\begin{aligned}
C_{XY}(t, t+\tau) &= E[(X(t) - E[X(t)])\,(Y(t+\tau) - E[Y(t+\tau)])] \\
&= E[X(t)Y(t+\tau) - E[X(t)]Y(t+\tau) - X(t)E[Y(t+\tau)] + E[X(t)]E[Y(t+\tau)]]
\end{aligned}
$$

since expectation is a linear operator,

$$
\begin{aligned}
C_{XY}(t, t+\tau) &= E[X(t)Y(t+\tau)] - E[X(t)]E[Y(t+\tau)] \\
&\quad - E[X(t)]E[Y(t+\tau)] + E[X(t)]E[Y(t+\tau)] \\
&= E[X(t)Y(t+\tau)] - E[X(t)]E[Y(t+\tau)] \\
&= R_{XY}(t, t+\tau) - E[X(t)]E[Y(t+\tau)]
\end{aligned}
$$

Properties of Covariance functions

(i) If $X(t)$ and $Y(t)$ are jointly WSS random processes, then

$$C_{XX}(\tau) = R_{XX}(\tau)\mu_X^2$$

$$C_{XY}(\tau) = R_{XY}(\tau)\mu_X\mu_Y$$

(ii) The variance of a WSS random process can be calculated as;

$$Var[X(t)] = \sigma_X^2 = E[(X(t) - E[X(t)])^2]$$
$$= R_{XX}(0) - \mu_X^2$$

(iii) Two random processes are called *uncorrelated* if their cross-covariance is zero;

$$C_{XY}(\tau) = 0$$

which leads us to

$$R_{XY}(t, t+\tau) = E[X(t)]E[Y(t+\tau)]$$

Therefore, it can be stated that if $X(t)$ and $Y(t)$ are statistically independent random processes, then they are uncorrelated.

(iv) In general, the correlation coefficient considering the random processes $X(t)$ and $Y(t)$ can be given as;

$$\rho_{XY}(t, t+\tau) = \frac{C_{XY}(t, t+\tau)}{\sigma_X(t)\sigma_Y(t)}$$

(v) If $X(t)$ and $Y(t)$ are jointly WSS random processes;

$$\rho_{XY}(\tau) = \frac{C_{XY}(\tau)}{\sigma_X\sigma_Y}$$
$$= \frac{R_{XX}(\tau) - \mu_X\mu_Y}{\sqrt{R_{XX}(0) - \mu_X^2}\sqrt{R_{YY}(0) - \mu_Y^2}}$$

Random processes can be classified simply as shown in Fig. 6.1.

Example 6.2 Consider an ergodic stationary random process having the autocorrelation function, $R_{XX}(\tau)$, as shown in Fig. 6.2.

(a) Derive the expected value of $X(t)$
(b) Derive the variance of $X(t)$

Fig. 6.1 Classification of random processes

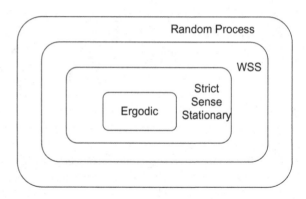

Fig. 6.2 $R_{XX}(\tau)$ to Example 6.2

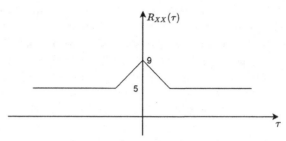

Solution 6.2 (a)

$$E[X(t)] = \lim_{\tau \to \infty} R_{XX}(\tau) = 5$$

(b)

$$
\begin{aligned}
Var[X(t)] &= (E[X(t)])^2 - E[X^2(t)] \\
&= 5^2 - R_{XX}(0) \\
&= 25 - 9 \\
&= 16
\end{aligned}
$$

6.4 Gaussian Random Process

Let $X(t)$ be a continuous random process. Consider related random variables such that $X_1 = X(t_1)$, $X_2 = X(t_2)$, ..., $X_N = X(t_N)$ at certain times from t_1 to t_N. A random vector contains the corresponding random variables

$$\mathbf{X} = \begin{bmatrix} X_1 \\ X_2 \\ \dots \\ X_N \end{bmatrix}$$

Each random variable of $X(t)$ have the following expected values shown in a matrix form;

$$\mu_{\mathbf{X}} = \begin{bmatrix} \mu_{X_1} \\ \mu_{X_2} \\ \cdots \\ \mu_{X_N} \end{bmatrix}$$

If these random variables are jointly Gaussian distributed, the corresponding joint probability density function can be defined as;

$$f_X(\mathbf{X}) = \frac{1}{\sqrt{(2\pi)^N det(\mathbf{C}_{\mathbf{XX}})}} e^{-\frac{1}{2}(\mathbf{X}-\mu_{\mathbf{X}})^T \mathbf{C}_{\mathbf{XX}}^{-1}(\mathbf{X}-\mu_{\mathbf{X}})}$$

where $\mathbf{C}_{\mathbf{XX}}$ is the covariance matrix such that its entries can be defined as;

$$\mathbf{C}_{\mathbf{XX}}(i, j) = \mathbf{C}_{\mathbf{XX}}(t_i, t_j) = E[(X_i - E[X_i])(X_j - E[X_j])]$$
$$= R_{XX}(t_i, t_j) - E[X(t_i)]E[X(t_j)]$$

Therefore;

A Gaussian random process can be characterized by knowing its autocorrelation function, $R_{XX}(t_i, t_j)$, and its mean matrix, $\mu_{\mathbf{X}}$.

If a Gaussian random process is WSS, then its mean is going to be a constant.

$$\mu_{X_i} = E[X(t_i)] = \mu_X = \text{constant}$$

Furthermore, under being WSS process condition, the autocorrelation and autocovariance funtions will be depend on the time shift $\tau = t_j - t_i$, such that;

$$R_{XX}(t_i, t_j) = R_{XX}(t_j - t_i)$$
$$C_{XX}(t_i, t_j) = C_{XX}(t_j - t_i)$$

Example 6.3 A WSS Gaussian random process having the mean values of 6, and the following autocorrelation function;

$$R_{XX}(\tau) = 100e^{-|\tau|}$$

Considering four random variables $X(t_1)$, $X(t_2)$, $X(t_3)$, and $X(t_4)$ given where $\tau = t_j - t_i$

(a) Find the variance of $X(t)$
(b) Derive the covariance matrix, $\mathbf{C}_{\mathbf{XX}}(\tau)$

Solution 6.3 (a)
$$Var[X(t)] = R_{XX}(0) - \mu_X^2$$
$$= 100 - 36$$
$$= 64$$

(b)

$$C_{XX} = \begin{bmatrix} R_{XX}(0) - \mu_X^2 & R_{XX}(1) - \mu_X^2 & R_{XX}(2) - \mu_X^2 & R_{XX}(3) - \mu_X^2 \\ R_{XX}(-1) - \mu_X^2 & R_{XX}(0) - \mu_X^2 & R_{XX}(1) - \mu_X^2 & R_{XX}(2) - \mu_X^2 \\ R_{XX}(-2) - \mu_X^2 & R_{XX}(-1) - \mu_X^2 & R_{XX}(0) - \mu_X^2 & R_{XX}(1) - \mu_X^2 \\ R_{XX}(-3) - \mu_X^2 & R_{XX}(-2) - \mu_X^2 & R_{XX}(-1) - \mu_X^2 & R_{XX}(0) - \mu_X^2 \end{bmatrix}$$

$$C_{XX} = \begin{bmatrix} 64 & 0.7879 & -22.466 & -31.021 \\ 0.7879 & 64 & 0.7879 & -22.466 \\ -22.466 & 0.7879 & 64 & 0.7879 \\ -31.021 & -22.466 & 0.7879 & 64 \end{bmatrix}$$

Example 6.4 Consider a Gaussian random process having mean and covariance functions as;
$$\mu_X(t) = 2t$$
$$C_{XX}(\tau) = 25e^{-\frac{\tau}{4}}$$

(a) Find $E[X(t = 1)]$
(b) Find $Var[X(t = 1)]$
(c) Find $E[Y(t)]$, where, $Y(t) = X(t) + X(3t)$
(d) Find $Var[Y(t = 1)]$
(e) Calculate the probability of $P(X(t = 1) < 1)$

Solution 6.4 (a)
$$E[X(t = 1)] = \mu_X(t = 1) = 2$$

(b)
$$Var[X(t = 1)] = C_{XX}(1, 1) = C_{XX}(0) = 25$$

(c)
$$E[Y(t)] = E[X(t) + X(3(t)]$$
$$= E[X(t)] + E[X(3t)]$$
$$= 2t + 6t$$
$$= 8t$$

(d)

$$
\begin{aligned}
Var[Y(t = 1)] &= Var[X(1) + X(3)] \\
&= Var[X(1)] + Cov[X(1)X(3)] + Cov[X(3)X(1)] + Var[X(3)] \\
&= C_{XX}(0) + C_{XX}(2) + C_{XX}(2) + C_{XX}(0) \\
&= 25 + 25e^{-\frac{2}{4}} + 25e^{-\frac{2}{4}} + 25 \\
&= 80.33
\end{aligned}
$$

(e)

$$
\begin{aligned}
P(X(t = 1) < 1) &= P\left(\frac{X(1) - \mu_X(1)}{\sigma_X(1)} < \frac{1 - \mu_X(1)}{\sigma_X(1)}\right) \\
&= P\left(\frac{X(1) - 2}{5} < \frac{1 - 2}{5}\right) \\
&= P\left(\frac{X(1) - 2}{5} < \frac{-1}{5}\right) \\
&= 1 - Q(-1/5) = Q(1/5) \quad \text{(from the normalized Gaussian distribution table)} \\
&= 0.5793
\end{aligned}
$$

6.5 Poisson Random Process

Poisson random process is a significant tool in order to model some physical phenomenon such as those involving counting events. For instance, a queueing system, a failure detection of data packages sent in a communication tool are well-known applications of Poisson process.

Let $X(t)$ be a random process representing the number of event happening in a time slot of $(0, t]$. Obviously, $X(t)$ is a non-decreasing, nonnegative integer-valued, and continuous time random process. The following assumptions have to be satisfied in order to build up a Poisson process;

(i) Let the time slot $(0, t]$ is divided into k subintervals of $\Delta t = \frac{t}{k}$. More than one event cannot be happened within a subinterval time of Δt.
(ii) Events are independent considering each subintervals. Hence the process is *memoryless*.

If the rate of events per unit time is notated as λ, then the expected number of events within $(0, t]$ is λt. Finally, the random process $X(t)$ can be modelled based on Poisson distribution such that;

$$
P(X(t) = k) = \frac{(\lambda t)^k}{k!} e^{-\lambda t} \quad \text{for } k = 0, 1, 2, \ldots
$$

The mean and variance functions of a Poisson process are;

$$
\mu_X(t) = E[X(t)] = \lambda t
$$
$$
Var[X(t)] = \lambda t
$$

Sum of Poisson Processes:
Let $X(t)$ and $Y(t)$ be two statistically independent Poisson processes having the parameters $\lambda_1 t$, and $\lambda_2 t$, respectively. A new random process is defined as $Z(t) = X(t) + Y(t)$ is also a Poisson process with the parameter of $(\lambda_1 + \lambda_2)t$

6.5.1 Autocorrelation Function of a Poisson Process

The expected number of a Poisson process within a time slot $(0, t]$ is $E[X(t)] = \lambda t$. In addition, the variance of a Poisson distribution can be defined as;

$$
\begin{aligned}
Var[X(t)] &= E[(X(t) - E[X(t)])^2] \\
&= E[X^2(t) - 2X(t)E[X(t)] + (E[X(t)])^2] \\
&\text{since } E[.] \text{is a linear operator} \\
&= E[X^2(t)] - 2\,(E[X(t)])^2 + (E[X(t)])^2 \\
&= E[X^2(t)] - (E[X(t)])^2 \\
\lambda t &= E[X^2(t)] - (\lambda t)^2
\end{aligned}
$$

Thereafter, considering a Poisson process, the average power of the signal is derived as;

$$
E[X^2(t)] = \lambda t + (\lambda t)^2
$$

In order to calculate the autocorrelation function, $R_{XX}(t_1, t_2)$, of a Poisson process;

$$
\begin{aligned}
R_{XX}(t_1, t_2) &= E[X(t_1)Xt_2)] \\
&\textbf{if } t_2 > t_1 \\
&= E[X(t_1)(X(t_1) + X(t_2) - X(t_1))] \\
&= E[X^2(t_1)]E[(X(t_1)(X(t_2) - X(t_1))] \\
&= \lambda t_1 + (\lambda t_1)^2 + \lambda t_1 \lambda(t_2 - t_1) \\
&= \lambda t_1 + \lambda^2 t_1 t_2 \\
&\textbf{if } t_2 < t_1 \\
&= E[X(t_2)(X(t_2) + X(t_1) - X(t_2))] \\
&= E[X^2(t_2)]E[(X(t_2)(X(t_1) - X(t_2))] \\
&= \lambda t_2 + (\lambda t_2)^2 + \lambda t_2 \lambda(t_1 - t_2) \\
&= \lambda t_2 + \lambda^2 t_1 t_2
\end{aligned}
$$

$$R_{XX}(t_1, t_2) = \begin{cases} \lambda t_1 + \lambda^2 t_1 t_2 & \text{if } t_2 \geq t_1 \\ \lambda t_2 + \lambda^2 t_1 t_2 & \text{if } t_2 \leq t_1 \end{cases}$$

Therefore, the Poisson process is not a WSS process.

6.5.2 Covariance Function of a Poisson Process

The covariance function of a Poisson process can be written as;

$$\begin{aligned} C_{XX}(t_1, t_2) &= R_{XX}(t_1, t_2) - E[X(t_1)X(t_2)] \\ &= \lambda t_1 + \lambda^2 t_1 t_2 - \lambda t_1 \lambda t_2 = \lambda t_1 \quad \text{if } t_2 \geq t_1 \\ &= \lambda t_2 + \lambda^2 t_1 t_2 - \lambda t_2 \lambda t_1 = \lambda t_2 \quad \text{if } t_1 \geq t_2 \end{aligned}$$

Therefore, the covariance function is simply defined as;

$$C_{XX}(t_1, t_2) = \lambda min(t_1, t_2)$$

6.6 Problems

6.1 Let $X(t)$ be a WSS random process with

$$E[X(t)] = 4$$
$$R_{XX}(\tau) = 25 + 16e^{-|\tau|}$$

If a new random process is defined as;

$$Y(t) = \int_{-\pi}^{\pi} 2X(t)dt$$

(a) Derive $E[Y(t)]$
(b) Calculate Var[Y(t)]

6.2 Consider a Gaussian random process having zero mean, and the following auto-correlation function

Fig. 6.3 $R_{XX}(\tau)$ to Problem 6.3

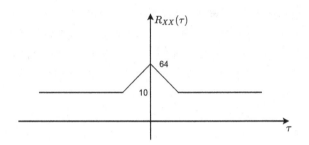

$$R_{XX}(\tau) = \frac{sin(\pi\tau)}{\pi\tau}$$

Construct the covariance matrix for $\tau = 0, 1, 2, 3, 4$

6.3 Consider an ergodic stationary random process having the autocorrelation function, $R_{XX}(\tau)$, as shown in Fig. 6.3

(a) Derive the expected value of $X(t)$
(b) Derive the variance of $X(t)$

6.4 Let $X(t)$ be a WSS random process defined as;

$$X(t) = \Theta$$

where Θ is a continuous random variable having a uniform probability distribution between $-\pi < \Theta < \pi$;

$$f_\Theta(\theta) = \begin{cases} \frac{1}{2\pi} & -\pi < \theta < \pi \\ 0 & elsewhere \end{cases}$$

(a) Calculate $\mu_X(t)$
(b) Derive thew autocorrelation function, $R_{XX}(t_1, t_2)$
(c) Is, $X(t)$ a WSS process?

6.5 Considering two statistically independent random processes, $X(t)$, and $Y(t)$, then define a new random process $Z(t) = X(t)Y(t)$;

(a) Derive $\mu_Z(t)$ in terms of $\mu_X(t)$ and $\mu_Y(t)$
(b) Find $R_{ZZ}(t_1, t_2)$ in terms of $R_{XX}(t_1, t_2)$ and $R_{YY}(t_1, t_2)$

6.6 Consider a WSS random process $X(t)$ with

$$E[X(t)] = 2$$
$$R_{XX}(\tau) = e^{-2\pi|\tau|}$$

If $Y(t) = 2X(t)$;

(a) Find $\mu_Y(t)$
(b) Define $R_{XY}(\tau)$
(c) Define $R_{YY}(\tau)$
(d) Is $Y(t)$ a WSS process?

6.7 Repeat Problem 6 for

$$Y(t) = 2X(t) + 1$$

6.8 Considering the following covariance matrix of two Gaussian random processes

$$\mathbf{C_{XX}}(\tau) = \begin{bmatrix} 9 & 0 \\ 0 & 16 \end{bmatrix}$$

Discuss the following statements whether they are correct or not,

(a) $Var[X_1(t)] = 9$ and $Var[X_2(t)] = 16$
(b) $X_1(t)$ and $X_2(t)$ are uncorrelated.
(c) $X_1(t)$ and $X_2(t)$ are orthogonal.

6.9 Let $X(t)$ be a WSS random process with

$$R_{XX}(\tau) = e^{-2\pi \tau^2}$$

and $Y(t) = \frac{d(X(t))}{dt}$

(a) Show that $\mu_Y(t) = 0$
(b) Derive $R_{XY}(\tau)$
(c) Derive $R_{YY}(\tau)$
(d) Is $Y(t)$ a WSS process?

6.10 Let $X(t)$ be a WSS random process with

$$R_{XX}(\tau) = e^{-|\tau|}$$
$$\mu_X(t) = 3$$

a new random process

$$Y(t) = X(t)\cos(wt + \Theta)$$

where Θ is a uniformly distributed random variable in $-\pi < \theta < \pi$

(a) Find $\mu_Y(t) = 0$
(b) Derive $R_{YY}(\tau)$
(c) Is $Y(t)$ a WSS process?

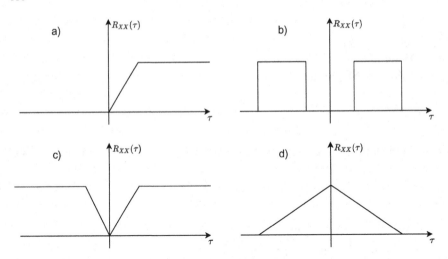

Fig. 6.4 Problem 6.12

6.11 Repeat Problem 10 if;

$$Y(t) = X(t)sin(wt + \Theta)$$

where Θ is a uniformly distributed random variable in $-\frac{\pi}{2} < \theta < \frac{\pi}{2}$

6.12 Which one of the following figures can represent a valid $R_{XX}(\tau)$ (Fig. 6.4);

Chapter 7
Spectral Analysis of Random Processes

This chapter is dedicated to give a brief introduction to the frequency domain analysis of random processes. The Fourier transform transfers a function $f(t)$ from the time domain into the frequency domain. Please note that, $f(t)$ should satisfy the Dirichlet conditions in order to have its Fourier transform. The Fourier transform of a function $f(t)$ is;

$$\mathcal{F}\{f(t)\} = F(jw)$$
$$= \int_{-\infty}^{\infty} f(t)e^{-jwt}\,\mathrm{d}t$$

where $j = \sqrt{-1}$. The function of $F(jw)$; is called the spectrum of $f(t)$ within the concept of the probability theory. If $F(jw)$ is known, then $f(t)$ can be derived using inverse Fourier transform such that;

$$f(t) = \mathcal{F}^{-1}\{F(jw)\}$$
$$= \frac{1}{2\pi} \int_{-\infty}^{\infty} F(jw)e^{jwt}\,\mathrm{d}w$$

A detailed knowledge about the Fourier transform are supplied in Appendix section.

© The Author(s), under exclusive license to Springer Nature Switzerland AG 2022   117
M. Catak et al., *Probability and Random Variables for Electrical Engineering*,
Studies in Systems, Decision and Control 390,
https://doi.org/10.1007/978-3-030-82922-3_7

7.1 The Power Density Spectrum

7.1.1 Continuous Time Random Processes

Let $x(t)$ be an appropriate realization of a continuous time random process. In order to hold Dirichlet conditions, a part of $x(t)$ within a finite period of T is obtained by means of a windowing function $w(t)$, such that;

$$w(t) = \begin{cases} 1 & \text{for } \frac{-T}{2} < t < \frac{T}{2} \\ 0 & \text{elsewhere} \end{cases}$$

then;

$$x_T(t) = w(t)x(t)$$

The Fourier transform of $x_T(t)$ is

$$X_T(jw) = \int_{-\frac{T}{2}}^{\frac{T}{2}} x_T(t)e^{-jwt}dt$$

$$= \int_{-\frac{T}{2}}^{\frac{T}{T}} x(t)e^{-jwt}dt$$

If $x(t)$ is the potential difference across a 1 Ohm resistor with unit of Volt, the unit of $X_T(jw)$ is Volt per Hertz. The average power is calculated taking into account T of period as;

$$P(T) = \frac{1}{T}\int_{-\frac{T}{2}}^{\frac{T}{2}} x^2(t)dt$$

The Parseval's theorem states that;

$$\int_{-\frac{T}{2}}^{\frac{T}{2}} x^2(t)dt = \frac{1}{2\pi}\int_{-\infty}^{\infty} |X_T(jw)|^2\,dw$$

Thereafter, the average power is written in terms of $X_T(jw)$ as;

$$P(T) = \frac{1}{T}\frac{1}{2\pi}\int_{-\infty}^{\infty} |X_T(jw)|^2\,dw$$

Considering that a random process, the expected value of the average power over its period of T is;

$$P_{XX} = \lim_{T \to \infty} \frac{1}{T} \int_{-\frac{T}{2}}^{\frac{T}{2}} E\left[x^2(t)\right] dt$$

$$= \frac{1}{2\pi} \int_{-\infty}^{\infty} \lim_{T \to \infty} \frac{E\left[|X_T(jw)|^2\right]}{T} dw$$

Finally, the power density spectrum, $S_{XX}(jw)$, is defined as;

$$S_{XX}(jw) = \lim_{T \to \infty} \frac{E\left[|X_T(jw)|^2\right]}{T}$$

Then, the expected value of the average power is expressed by means of the power density spectrum ;

$$P_{XX} = \frac{1}{2\pi} \int_{-\infty}^{\infty} S_{XX}(jw) dw.$$

7.2 Properties of the Power Density Spectrum

The following properties of the power density spectrum, $S_{XX}(jw)$, help us to understand and to analyze a random process in the frequency domain.

Properties of $S_{XX}(jw)$
(1) $S_{XX}(jw)$ is a real and nonnegative function. One can easily infer that the expected value of the square of an magnitude function must be nonnegative and real.
(2) If $X(t)$ is a real function, then $S_{XX}(jw)$ is an even function such that;

$$S_{XX}(-jw) = S_{XX}(jw)$$

(3) For a WSS random process, the autocorrelation function and the power spectral density function are Fourier transform pairs of each other;

$$S_{XX}(jw) = \int_{-\infty}^{\infty} R_{XX}(\tau)e^{-jw\tau} d\tau$$

$$R_{XX}(\tau) = \frac{1}{2\pi} \int_{-\infty}^{\infty} S_{XX}(jw)e^{jw\tau} dw$$

(4) The expected value of the average power of a signal is

$$P_{XX} = \frac{1}{2\pi} \int_{-\infty}^{\infty} S_{XX}(jw)dw$$
$$= R_{XX}(0)$$

(5) Considering a WSS random process,

$$C_{XX}(\tau) = R_{XX}(\tau) - \mu_X^2$$
therefore
$$R_{XX}(\tau) = C_{XX}(\tau) + \mu_X^2$$
since Fourier transform is an linear operator
$$\mathcal{F}\{R_{XX}(\tau)\} = \mathcal{F}\{C_{XX}(\tau) + \mu_X^2\}$$
$$S_{XX}(jw) = \mathcal{F}\{C_{XX}(\tau)\} + 2\pi \mu_X^2 \delta(w)$$

(6) Considering a WSS random process, the cross power spectral density can be derived as;

$$S_{XY}(jw) = \mathcal{F}\{R_{XY}(\tau)\}$$
$$= \int_{-\infty}^{\infty} R_{XY}(\tau)e^{-jw\tau}d\tau$$

Example 7.1 Let $X(t)$ and $Y(t)$ be two jointly WSS random processes. A new random process is defined as $Z(t) = X(t) + Y(t)$

(a) Derive the autocorrelation function of $Z(t)$ in terms of $R_{XX}(\tau)$ and $R_{YY}(\tau)$
(b) Derive the power density spectrum of $Z(t)$

Solution 7.1 (a)

$$R_{ZZ}(\tau) = E[Z(t)Z(t+\tau)]$$
$$= E[(X(t)Y(t))(X(t+\tau)Y(t+\tau))]$$
$$= E[X(t)X(t+\tau)] + E[X(t)Y(t+\tau)] + E[Y(t)X(t+\tau)] + E[Y(t)Y(t+\tau)]$$
$$= R_{XX}(\tau) + R_{XY}(\tau) + R_{YX}(\tau) + R_{YY}(\tau)$$

(b) Since the power density spectrum and the autocorrelation function are Fourier transform pairs;

$$S_{ZZ}(j\omega) = \mathcal{F}\{R_{ZZ}(\tau)\}$$
$$= \mathcal{F}\{R_{XX}(\tau) + R_{XY}(\tau) + R_{YX}(\tau) + R_{YY}(\tau)\}$$
$$= S_{XX}(j\omega) + S_{XY}(j\omega) + S_{YX}(j\omega) + S_{YY}(j\omega)$$

Example 7.2 Let $X(t)$ be a WSS random process with the following autocorrelation function

$$R_{XX}(\tau) = e^{-|\tau|}$$

(a) Find $S_{XX}(j\omega)$
(b) Calculate the average power of a sample.

Solution 7.2 (a)

$$S_{XX}(j\omega) = \int_{-\infty}^{\infty} R_{XX}(\tau)e^{-j\omega\tau}d\tau$$

$$= \int_{-\infty}^{0} e^{\tau}e^{-j\omega\tau}d\tau + \int_{0}^{\infty} e^{-\tau}e^{-j\omega\tau}d\tau$$

$$= \int_{-\infty}^{0} e^{\tau(1-j\omega)}d\tau + \int_{0}^{\infty} e^{-\tau(1+j\omega)}d\tau$$

$$= \frac{1}{1-j\omega}e^{\tau(1-j\omega)}\Big|_{\tau=-\infty}^{0} \ - \ \frac{1}{1+j\omega}e^{-\tau(1+j\omega)}\Big|_{\tau=0}^{\infty}$$

$$= \frac{1}{1-j\omega} + \frac{1}{1+j\omega}$$

$$= \frac{1+jw+1-j\omega}{1-j^2\omega^2}$$

$$= \frac{2}{1+\omega^2}$$

(b) The average power;

$$P_{XX} = \frac{1}{2\pi}\int_{-\infty}^{\infty} S_{xx}(j\omega)d\omega$$

$$= \frac{1}{2\pi}\int_{-\infty}^{\infty} \frac{2}{1+w^2}d\omega$$

$$= \frac{1}{2\pi}\left(2\tan^{-1}(\omega)\Big|_{w=-\infty}^{\infty}\right)$$

$$= \frac{1}{2\pi}\left(2(\frac{\pi}{2}+\frac{\pi}{2})\right)$$

$$= 1$$

or, considering in time domain

$$P_{XX} = E[X^2(t)]$$
$$= R_{XX}(0)$$
$$= e^{-0}$$
$$= 1.$$

7.3   Discrete Time Random Processes

Suppose that $X[k]$ is a discrete time WSS random process having $E[X[k]] = \mu_X$, and the autocorrelation function of $R_{XX}[n]$. The power spectral density of $X[k]$ can be defined using Discrete Time Fourier Transform (DTFT) as;

$$S_{XX}(e^{jw}) = \sum_{n=-\infty}^{\infty} R_{XX}[n]e^{-jwn}$$

By definition, $S_{XX}(e^{jw})$ is a periodic function with the period of 2π. Therefore, the range of the frequency must be $-\frac{\pi}{2} < w \le \frac{\pi}{2}$. As shown in continuous time processes; the autocorrelation function and the power spectral density are DTFT pairs;

$$R_{XX}[n] = \frac{1}{2\pi} \int_{-\pi}^{\pi} S_{XX}(e^{jw})e^{jwn}dw.$$

7.4   Problems

7.1 Let $X(t)$ be a WSS random process having

$$R_{XX}(\tau) = e^{-|\tau|}$$

(a) Derive $S_{XX}(j\omega)$
(b) If $Y(t) = X(3t)$, derive $S_{YY}(j\omega)$
(c) Find $R_{YY}(\tau)$ using the result obtained in part (b)

7.2 Repeat Problem 7.1 if
$$Y(t) = X(-3t)$$

7.3 Let $X(t)$ be a WSS random process with

$$S_{XX}(j\omega) = \frac{8}{16 + \omega^2}$$

(a) Derive $R_{XX}(\tau)$
(b) If $Y(t) = \frac{d}{dt}X(t)$, find $S_{XY}(j\omega)$
(c) Find $S_{YY}(j\omega)$
(d) Find $R_{XY}(\tau)$
(e) Find $R_{YY}(\tau)$

7.4 Repeat Problem 7.3 if

$$S_{XX}(j\omega) = \frac{1}{25 + \omega^2}$$

7.5 The autocorrelation function of a WSS random process is

$$R_{XX}(\tau) = 4 + e^{-\frac{|\tau|}{2}}$$

(a) Find $S_{XX}(j\omega)$
(b) Calculate the average power of a sample
(c) Calculate the power of the signal within the frequency band of $\left(-\frac{\pi}{4} < w \leqslant \frac{\pi}{4}\right)$

7.6 Consider a WSS random process $X(t)$ having its power spectral density function as $S_{XX}(jw)$. If

$$S_{YY}(j\omega) = \frac{d^2 S_X X(jw)}{dw^2}$$

Derive $R_{YY}(\tau)$ in terms of $R_{XX}(\tau)$

7.7 The autocorrelation function of a WSS random process is given as

$$R_{XX}(\tau) = \frac{5}{1 + \tau^2}$$

(a) Find $S_{XX}(j\omega)$
(b) Calculate the average power of a sample

7.8 Considering a discrete random sequence $X[k]$ with zero mean

(a) Show that

$$R_{XX}[n] = \sigma_X^2 \delta[n]$$

(b) Using part (a) derive that

$$S_{XX}(e^{jw}) = \sigma_X^2$$

7.9 Considering the discrete random sequence $X[k]$ given in Problem 7.8, if

$$Y[k] = X[k - 2]$$

(a) Derive $R_{YY}[n]$ in terms of $R_{XX}[n]$
(b) Calculate $S_{YY}(e^{jw})$ in terms of $S_{XX}(e^{jw})$

7.10 The autocorrelation function of a WSS random sequence is given as;

$$R_{XX}[n] = 2\left(\frac{1}{2}\right)^{|}n|$$

(a) Find $S_{XX}(e^{jw})$
(b) Calculate the average power of a sample

7.11 Considering the autocorrelation function of a WSS random sequence;

$$R_{XX}[n] = 4\cos[w_0 n]$$

where w_0 is a constant.

(a) Derive $S_{XX}(e^{jw})$
(b) Calculate the average power of a sample.

Chapter 8
Linear Systems with Random Inputs

In the realm of the signals and system theory, any system consists of three parts, such that the input (denoted as $x(t)$, the impulse response $(h(t))$ which characterizes the process, and the output $(y(t))$ as shown in Fig. 8.1.

Let T be an operator ruling the input $x(t)$ to the output $y(t)$

$$y(t) = T\{x(t)\}$$

If the input is a unit impulse function, i.e. $\delta(t)$, then the corresponding output is called the impulse response; by definition

$$h(t) = T\{\delta(t)\}.$$

8.1 System Properties

8.1.1 Linear Systems

A system is called linear if and only if the ruling operator $T\{.\}$ has the following properties:

(i) $T\{\alpha x(t)\} = \alpha y(t)$, where α is a constant
(ii) $T\{x_1(t) + x_2(t)\} = T\{x_1(t)\} + T\{x_2(t)\}$

These are called *homogeneity* and *additivity* properties, respectively. Consequently, if these two properties generalized and combined into one definition;

M. Catak et al., *Probability and Random Variables for Electrical Engineering*,
Studies in Systems, Decision and Control 390,
https://doi.org/10.1007/978-3-030-82922-3_8

Fig. 8.1 A basic system
representation

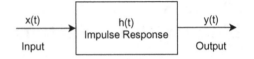

$$y(t) = T \left\{ \sum_{i=1}^{\infty} \alpha_i x_i(t) \right\} = \sum_{i=1}^{\infty} \alpha_i T \{x_i(t)\} = \sum_{i=1}^{\infty} \alpha_i y_i(t).$$

8.1.2 Time-Invariant Systems

A system is time-invariant if any time shift in the input signal results in the same
time shift in the output signal, such that;

$$y(t) = T\{x(t)\}$$
$$y(t - t_0) = T\{x(t - t_0)\}$$

Otherwise, the system is called time-varying.

8.1.3 Stable Systems

In a general point of view, if a system returns its equilibrium points after any inter-
ruption by external source, such system is called a stable system.

If the input of a system is bounded, $|x(t)| < K_1$, where K_1 is a finite number; the
corresponding output $y(t)$ is also bounded if and only if the impulse response of the
system $h(t)$ satisfies the following condition;

$$\int_{-\infty}^{\infty} |h(t)| dt < \infty$$

Such system is called bounded input bounded output (BIBO) stable.

8.1.4 Causal Systems

A system is said to be causal if the output of the system $y(t)$ at an arbitrary time t_0
depends only on the input at and or before the time t_0. In other words, the present of

output can depend on present and/or past inputs. This definition is hold only if the impulse response of the system $h(t)$ satisfies;

$$h(t) = 0 \quad \forall \quad t < 0$$

In literature, the causal systems are also known as physically realizable systems.

8.1.5 Linear Time-Invariant Systems

In this section, the expression of the output of a LTI system in terms of the input $x(t)$ and the impulse response $h(t)$ will be discussed. Any continuous time function can be represented by using unit impulse function;

$$x(t) = \int_{-\infty}^{\infty} x(\tau)\delta(t - \tau)d\tau$$

Thereafter, the output of such a system can be defined as;

$$\begin{aligned}
y(t) &= T\{x(t)\} \\
&= T\left\{ \int_{-\infty}^{\infty} x(\tau)\delta(t - \tau)d\tau \right\} \\
&= \int_{-\infty}^{\infty} x(\tau)T\{\delta(t - \tau)\}d\tau
\end{aligned}$$

If the system is linear (no information about the time-invariance of the system), then the impulse response can be expressed as;

$$T\{\delta(t - \tau)\} = h(t, \tau)$$

hence, the output $y(t)$ can be derived as;

$$y(t) = \int_{-\infty}^{\infty} x(\tau)h(t, \tau)d\tau$$

which is called superposition theorem in signal processing.

If the system is linear and time-invariant (LTI), then

$$T\{\delta(t - \tau)\} = h(t - \tau)$$

hence, the output signal is represented as;

$$y(t) = \int_{-\infty}^{\infty} x(\tau)h(t-\tau)d\tau$$

which is called the convolution integral.

Example 8.1 Consider a system with the following input-output relation;

$$y(t) = 2x(t) + 1$$

Determine whether the system is

(a) Linear
(b) Stable
(c) Causal
(d) Time-invariant
(e) LTI

Solution 8.1 (a) Let assign two arbitrary input functions as $x_1(t)$ and $x_2(t)$

$$y_1(t) = 2x_1(t) + 1$$
$$y_2(t) = 2x_2(t) + 1$$
$$y(t) = 2(x_1(t) + x_2(t)) + 1$$

$y(t) \neq y_1(t) + y_2(t)$ therefore, the system is nonlinear.
(b) If the input is assign as $\delta(t)$, then the corresponding output will be $h(t)$;

$$h(t) = 2\delta(t) + 1$$

it will be checked whether the following statement is hold or not;

$$\int_{-\infty}^{\infty} |h(t)|dt < \infty$$

$$\int_{-\infty}^{\infty} |2\delta(t) + 1|dt \not< \infty$$

therefore, the system is unstable.
(c) Select an arbitrary time as t_0;

$$y(t_0) = 2x(t_0) + 1$$

since the output is only depended on t_0, then the system is causal.
(d) If the input is shifted in time with t_0, corresponding output will be $2x(t_0) + 1$, which equals the output shifted in time with t_0. Therefore, the system is time-invariant.
(e) The system is time-invariant but it is nonlinear, therefore the system is not LTI.

8.1.5.1 Frequency Response

The impulse response $h(t)$ of a linear time-invariant system represents the general behaviour of such a system by means of she convolution integral between $h(t)$ and any input $x(t)$. A similar representation of a LTI system can be achieved as well in frequency domain using Fourier transform.

$$Y(jw) = \int_{-\infty}^{\infty} y(t)e^{-jwt} dt$$

$$= \int_{-\infty}^{\infty} x(\tau) \left(\int_{-\infty}^{\infty} h(t-\tau)e^{-jw(t-\tau)} dt \right) e^{-jw\tau} d\tau$$

Then it can be written as;

$$Y(jw) = X(jw)H(jw)$$

where $X(jw)$ and $H(jw)$ are the Fourier transforms of $x(t)$, and $h(t)$, respectively. The term of $H(jw)$ is called the frequency response of the system. For some simple cases, for instance if the relation between the input signal and the output signal is defined by means of a linear constant coefficient differential equation, if the input is $x(t) = e^{jwt}$, then the output can be expressed as $y(t) = H(jw)e^{jw}$.

Example 8.2 Consider a stable, causal LTI system having the following input-output relation;

$$\frac{dy(t)}{dt} + 4y(t) = x(t)$$

Derive the transfer function of the system.

Solution 8.2 Since, if the input is $x(t) = e^{jwt}$, then the corresponding output will be $y(t) = H(jw)e^{jw}$. Thereafter;

$$\frac{dy(t)}{dt} + 4y(t) = x(t)$$

$$\frac{dH(jw)e^{jwt}}{dt} + 4H(jw)e^{jwt} = e^{jwt}$$

$$jwH(jw)e^{jwt} + 4H(jw)e^{jwt} = e^{jwt}$$

$$e^{jwt}(jwH(jw) + 4H(jw)) = e^{jwt}$$

$$H(jw) = \frac{1}{1 + j4w}$$

8.2 Systems with Random Inputs

As discussed above, the relationship between the input and the output of an LTI system can be expressed in convolution integral. Assuming with the impulse response of a LTI system is deterministic, the corresponding random output of an LTI system of a random input $X(t)$ is (Fig. 8.2);

$$Y(t) = \int_{-\infty}^{\infty} h(\tau)X(t - \tau)\mathrm{d}\tau$$

8.2.1 Statistical Analysis of Random Outputs

8.2.1.1 Expected Value of the Random Output, $E[Y(t)]$

By definition, the expected value of the random output is

$$
\begin{aligned}
E[Y(t)] &= E\left[\int_{-\infty}^{\infty} X(\tau)h(t - \tau)\mathrm{d}\tau\right] \\
&= E\left[\int_{-\infty}^{\infty} h(\tau)X(t - \tau)\mathrm{d}\tau\right] \\
&= \int_{-\infty}^{\infty} E[X(t - \tau)]h(\tau)\mathrm{d}\tau
\end{aligned}
$$

If $X(t)$ is a WSS process, then it must have a constant mean. Therefore, the expected value of the random output $Y(t)$ is;

$$E[Y(t)] = E[X(t)] \int_{-\infty}^{\infty} h(t)\mathrm{d}t$$

Since $\int_{-\infty}^{\infty} h(t)\mathrm{d}t$ is a finite constant for a stable system; then the expected value of a random output is constant for such stable systems having WSS random input.

Fig. 8.2 A basic random-input random-output system representation

8.2.1.2 Variance of the Random Output, $Var[Y(t)]$

It is appropriate to derive the mean-squared value of the random output before discussing the variance of the output.

$$E\left[Y^2(t)\right] = E\left[\int_{-\infty}^{\infty} h\left(\tau_1\right) X\left(t - \tau_1\right) d\tau_1 \int_{-\infty}^{\infty} h\left(\tau_2\right) X\left(t - \tau_2\right) d\tau_2\right]$$

Assuming that the expected value and the integral operators can be interchanged, then;

$$E\left[Y^2(t)\right] = \int_{-\infty}^{\infty}\int_{-\infty}^{\infty} E\left[X\left(t - \tau_1\right) X\left(t - \tau_2\right)\right] h\left(\tau_1\right) h\left(\tau_2\right) d\tau_1 d\tau_2$$

If the random input $X(t)$ is a WSS process;

$$E\left[X\left(t - \tau_1\right) X\left(t - \tau_2\right)\right] = R_{XX}\left(\tau_1 - \tau_2\right)$$

then;

$$E\left[Y^2(t)\right] = R_{XX}\left(\tau_1 - \tau_2\right) \int_{-\infty}^{\infty}\int_{-\infty}^{\infty} h\left(\tau_1\right) h\left(\tau_2\right) d\tau_1 d\tau_2$$

The variance of the random output can be expresses in terms of the expected value and the mean-squared value of the random output such that;

$$Var[Y(t)] = E\left[(Y(t) - E[Y(t)])^2\right]$$
$$= E\left[Y^2(t)\right] - (E[Y(t)])^2$$
$$= R_{XX}\left(\tau_1 - \tau_2\right) \int_{-\infty}^{\infty}\int_{-\infty}^{\infty} h\left(\tau_1\right) h\left(\tau_2\right) d\tau_1 d\tau_2 - \left(E[X(t)] \int_{-\infty}^{\infty} h(t)dt\right)^2$$

8.2.1.3 Autocorrelation Function of Random Output

The autocorrelation between $Y(t)$ and $T(t + \tau)$ is given as;

$$R_{YY}(t, t + \tau) = E[Y(t)Y(t + \tau)]$$
$$= E\left[\int_{-\infty}^{\infty}\int_{-\infty}^{\infty} h\left(\tau_1\right) X\left(t - \tau_1\right) d\tau_1 \int_{-\infty}^{\infty} h\left(\tau_2\right) X\left(t + \tau - \tau_2\right) d\tau_2\right]$$
$$= \int_{-\infty}^{\infty}\int E\left[X\left(t - \tau_1\right) X\left(t + \tau - \tau_2\right)\right] h\left(\tau_1\right) h\left(\tau_2\right) d\tau_1 d\tau_2$$

using the equality of

$$E\left[X\left(t-\tau_1\right)X\left(t+\tau-\tau_2\right)\right] = R_{XX}\left(\tau+\tau_1-\tau_2\right)$$

$$R_{YY}(t,t+\tau) = \int_{-\infty}^{\infty}\int_{-\infty}^{\infty} R_{XX}\left(\tau+\tau_1-\tau_2\right)h\left(\tau_1\right)h\left(\tau_2\right)d\tau_1 d\tau_2$$

According to these results,

> If the random input $X(t)$ is a WSS process, then the corresponding random output is also a WSS process. In other words;
> (i) the expected value of $Y(t)$, $E[Y(t)]$ is a constant
> (ii) the autocorrelation function of $Y(t)$, $R_{YY}(t,t+\tau)$ depends only on τ.

In addition, the autocorrelation function of the random output $Y(t)$ is two-fold convolution of the autocorrelation function of the random input $X(t)$ with the impulse response of the system, such that;

$$R_{YY}(\tau) = R_{XX}(\tau) * h(-\tau) * h(\tau)$$

8.2.1.4   Cross-Correlation Function Between Random Input and Random Output

The cross-correlation function between the random input and the random output is described as;

$$R_{XY}(t,t+\tau) = E[X(t)Y(t+\tau)]$$

$$= E\left[X(t)\int_{-\infty}^{\infty} h\left(\tau_1\right)X\left(t+\tau-\tau_1\right)d\tau_1\right]$$

$$= \int_{-\infty}^{\infty} E\left[X(t)X\left(t+\tau-\tau_1\right)\right]h\left(\tau_1\right)d\tau_1$$

If $X(t)$ is a WSS random process, then

$$R_{XY}(\tau) = \int_{-\infty}^{\infty} R_{XX}\left(\tau-\tau_1\right)h\left(\tau_1\right)d\tau_1$$

or, it can be written as;

$$R_{XY}(\tau) = R_{XX}(\tau) * h(\tau)$$

In a similar manner; one derives

$$R_{YX}(\tau) = R_{XX}(\tau) * h(-\tau)$$

Finally, the autocorrelation function of the random output can be defined in terms of the cross-correlation functions, such as;

$$R_{YY}(\tau) = R_{XY}(\tau) * h(-\tau) = R_{YX}(\tau) * h(\tau)$$

8.3 Power Density Spectrum of Random Outputs

In this section, the power density spectrum of the random output will be explained in terms of the random input and the frequency response of an LTI system. As mentioned in Chap. 6 the autocorrelation function of a random process and its power density spectrum are Fourier transform pairs, namely

$$S_{XX}(jw) = \int_{-\infty}^{\infty} R_{XX}(\tau)e^{-jw\tau}d\tau$$

$$R_{XX}(\tau) = \frac{1}{2\pi} \int_{-\infty}^{\infty} S_{XX}(jw)e^{jw\tau}dw$$

In addition, convolution and multiplication operators are dual operators both in time and frequency domains. In other words, the multiplication operator in time domain turns into convolution operator in frequency domain; the convolution operator in time domain turns into multiplication operator in frequency domain.

$$x(t)h(t) \leftrightarrow X(jw) * H(jw)$$
$$x(t) * h(t) \leftrightarrow X(jw)H(jw)$$

Since Fourier transform is a linear operator, taking Fourier transform of both sides of correlation equation

$$R_{YY}(\tau) = R_{XX}(\tau) * h(-\tau) * h(\tau)$$
$$\mathcal{F}\{R_{YY}(\tau)\} = \mathcal{F}\{R_{XX}(\tau) * h(-\tau) * h(\tau)\}$$
$$S_{YY}(jw) = S_{XX}(jw)H^*(jw)H(jw)$$

where $H^*(jw)$ is the complex conjugate of $H(jw)$, and

$$H^*(jw)H(jw) = |H(jw)|^2$$

Thereafter, the power spectral density of the random output can be defined as;

$$S_{YY}(jw) = |H(jw)|^2 S_{XX}(jw)$$

8.4   Noisy Inputs

8.4.1   White Noise

Let $N(t)$ be a *WSS* random process defining a noise signal. If the power density spectrum of $N(t)$, i.e. $S_{NN}(jw)$, is constant over frequency domain, then $N(t)$ is called white noise. Otherwise, $N(t)$ is said to be colored noise. The autocorrelation function of $N(t)$, i.e. $R_{NN}(\tau)$ can be calculated by means of inverse Fourier transform, such that;

$$R_{NN}(\tau) = \frac{1}{2\pi} \int_{-\infty}^{\infty} S_{NN}(jw)e^{jw\tau}dw$$

where $S_{NN}(jw) = \kappa$. Therefore, the autocorrelation function of $N(t)$ is derived as;

$$R_{NN}(\tau) = \kappa\delta(\tau)$$

However, in practical cases, a pure white noise is impossible, since its expected value of the average power goes to infinity;

$$P_{NN} = \frac{1}{2\pi} \int_{-\infty}^{\infty} S_{NN}(jw)dw \rightarrow \infty$$

Therefore, for physically realizable systems a low-pass type bounded with a critical frequency, ω_c, is assumed to represent the power density spectrum of a white noise, such that;

$$S_{NN}(jw) = \begin{cases} \kappa & |w| < w_c \\ 0 & \text{elsewhere} \end{cases}$$

The corresponding autocorrelation function is calculated as;

$$
\begin{aligned}
R_{NN}(\tau) &= \frac{1}{2\pi} \int_{-\infty}^{\infty} S_{NN}(jw)e^{jw\tau}dw \\
&= \frac{1}{2\pi} \int_{-w_c}^{w_c} \kappa e^{jw\tau}dw \\
&= \frac{1}{2\pi} \left[\frac{\kappa}{j\tau}e^{jw\tau} \Big|_{-w_c}^{w_c} \right] \\
&= \frac{1}{2\pi} \left[\frac{\kappa}{j\tau} \left(e^{jw_c\tau} - e^{-jw_c\tau} \right) \right] \\
&= \frac{\kappa}{\pi\tau} \left[\frac{1}{j2} \left(e^{jw_c T} - e^{-jw_c\tau} \right) \right]
\end{aligned}
$$

using Euler equality of

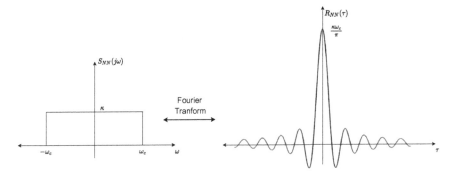

Fig. 8.3 Autocorrelation and Power spectrum density functions of a physically realizable white noise

$$\sin(w_c\tau) = \frac{1}{j2}\left(e^{jw_c\tau} - e^{-jw_c\tau}\right)$$

therefore,

$$R_{NN}(\tau) = \frac{\kappa}{\pi\tau}\sin(w_c\tau)$$

In the signal processing area, a well-known *Sinc* function is defined as;

$$\text{Sinc}(x) = \frac{\sin(nx)}{nx}$$

The autocorrelation function can be expressed in term of *Sinc* function as (Fig. 8.3);

$$R_{NN}(\tau) = \frac{\kappa w_c}{\pi}\text{Sinc}(w_c\tau)$$

8.5 Discrete Time Systems with Random Inputs

Considering a discrete time LTI system, the relation between input and the output can be given as;

$$y[n] = T\{x[n]\}$$
$$= T\left\{\sum_{k=-\infty}^{\infty} x[k]\delta[n-k]\right\}$$
$$= \sum_{k=-\infty}^{\infty} x[k]T\{\delta[n-k]\}$$
$$= \sum_{k=-\infty}^{\infty} x[k]h[n-k]$$

where $h[n]$ represents the impulse response of the corresponding system. Simply, the output can be derived by means of the convolution sum;

$$y[n] = x[n] * h[n]$$

$$= \sum_{k=-\infty}^{\infty} x[k]h[n-k]$$

Suppose that $X[k]$ is a discrete time WSS random process having $E[X[k]] = \mu_X$, and the autocorrelation function of $R_{XX}[n]$. The power spectral density of $X[k]$ can be defined using Discrete Time Fourier Transform (DTFT) as;

$$S_{XX}(e^{jw}) = \sum_{n=-\infty}^{\infty} R_{XX}[n]e^{-jwn}$$

and, the autocorrelation function and the power spectral density are DTFT pairs as discussed in Chap. 7;

$$R_{XX}[n] = \frac{1}{2\pi} \int_{-\pi}^{\pi} S_{XX}(e^{jw})e^{jwn}dw$$

The following equalities can be easily derived as discussed in continuous time case;

$$R_{XY}[n] = R_{XX}[n] * h[n]$$
$$R_{YX}[n] = R_{XX}[n] * h[-n]$$
$$R_{YY}[n] = R_{XX}[n] * h[n] * h[-n]$$
$$S_{YY}(e^{jw}) = |H(e^{jw})|^2 S_{XX}(e^{jw})$$

8.6 Problems

8.1 Consider a system with the following input-output relation;

$$y(t) = 2x(t) - x(t-1)$$

Determine whether the system is

(a) Linear
(b) Stable
(c) Causal
(d) Time-invariant
(e) LTI

8.2 Consider a stable, causal, and LTI system having the following input-output relation;

$$\frac{dy(t)}{dt} + 2y(t) = x(t)$$

Let $X(t)$ be a WSS random process employed as an input for such a system having the following autocorrelation function;

$$R_{XX}(\tau) = \delta(\tau)$$

(a) Derive the transfer function of the system.
(b) Find the expected value of $Y(t)$ in terms of $E[X(t]$
(c) Find $R_{YY}(\tau)$

8.3 A WSS random process $X(t)$ is employed as an input for such a system with

$$h(t) = te^{-2t}u(t)$$

(a) Derive the transfer function of the system, $H(jw)$.
(b) Find the expected value of $Y(t)$, if $E[X(t] = 5$

8.4 Repeat Problem 8.3, if

$$h(t) = (t-2)e^{-2(t-2)}u(t-2)$$

8.5 Consider the system given in Problem 8.4, if the autocorrelation function of the input is given as;

$$R_{XX}(\tau) = 5e^{-\frac{|\tau|}{3}}$$

(a) Derive $S_{XX}(jw)$
(b) Find $R_{YY}(\tau)$
(c) Derive $S_{YY}(jw)$
(d) Calculate the average power in $X(t)$
(e) Calculate the average power in $Y(t)$

8.6 Considering the circuit shown in Fig. 8.4 where $L = 1$ mH; and $R = 1$ kΩ.

(a) Derive the transfer function of the system, $H(jw)$.
    If $V_{in}(t)$ is represented by a WSS random process of $X(t)$ having autocorrelation function of

$$R_{XX}(\tau) = 6e^{-\frac{|\tau|}{2}}$$

(b) Find $E[X(t)]$
(c) Find $E[Y(t)]$
(d) Calculate $Var[Y(t)]$

Fig. 8.4 A simple RL circuit

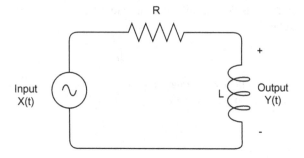

Fig. 8.5 A simple RC circuit

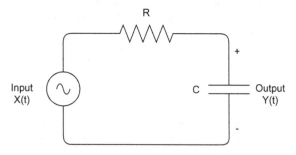

8.7 Consider the circuit given in Problem 8.6

(a) Derive $S_{XX}(jw)$
(b) Derive $S_{YY}(jw)$
(c) Find $R_{YY}(\tau)$
(d) Calculate the average power in $X(t)$
(e) Calculate the average power in $Y(t)$

8.8 Considering the circuit shown in Fig. 8.5 where $C = 1\ \mu F$; and $R = 2\ k\Omega$.

(a) Derive the transfer function of the system, $H(jw)$.
(b) Find the impulse response of the systems, $h(t)$.
      If $V_{in}(t)$ is represented by a WSS random process of $X(t)$ having autocorrelation
      function of
$$R_{XX}(\tau) = 3e^{-|\tau|}$$

(c) Find $E[X(t)]$
(d) Find $E[Y(t)]$
(e) Calculate $Var[Y(t)]$

8.9 Consider the circuit given in Problem 8.8

(a) Derive $S_{XX}(jw)$
(b) Derive $S_{YY}(jw)$
(c) Find $R_{YY}(\tau)$

Fig. 8.6 A simple RLC
circuit

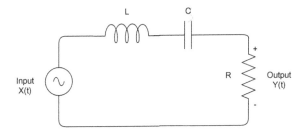

(d) Calculate the average power in $X(t)$
(e) Calculate the average power in $Y(t)$

8.10 Considering the circuit shown in Fig. 8.6 where $L = 1$ mH, $C = 1$ μF, and $R = 1$ kΩ.

(a) Derive the transfer function of the system, $H(jw)$.
(b) Sketch the transfer function of the system, $|H(jw)|$. If $V_{in}(t)$ is represented by a WSS random process of $X(t)$ having autocorrelation function of

$$R_{XX}(\tau) = 4e^{-2|\tau|}$$

(c) Find $E[X(t)]$
(d) Find $E[Y(t)]$
(e) Calculate $Var[Y(t)]$

8.11 Consider the circuit given in Problem 8.10

(a) Derive $S_{XX}(jw)$
(b) Derive $S_{YY}(jw)$
(c) Find $R_{YY}(\tau)$
(d) Calculate the average power in $X(t)$
(e) Calculate the average power in $Y(t)$

8.12 Consider a stable, causal, and LTI system having the following input-output relation;

$$y[n] = x[n] + y[n-1]$$

(a) Derive the impulse response, $h[n]$. If

$$R_{XX}[n] = (\frac{1}{2})^n u[n]$$

(b) Find $R_{YX}[n]$
(c) Find $R_{YY}[n]$
(d) Derive $S_{XX}(e^{jw})$
(e) Calculate $S_{YY}(e^{jw})$.

Chapter 9
Random Samples

Probability theory is an efficient tool in order to analyze physical systems including random nature. The randomness property could be arise due to our lack of knowledge about the process, or the process might have an intrinsic random nature. In general, without discussing the source of the random nature of the process, the experiments have to be done, then the corresponding data is obtained in discrete form. Therefore, sampling and sample distributions are significant topics in the probability theory area.

9.1 Random Sample Sequences

Assuming with the observed property of a system can be measurable, a random sample, $\mathbf{X}$, is defined as;

$$\mathbf{X} = [X_1, X_2, \ldots, X_n]$$

where $X_1, X_2, \ldots, X_n$ are statistically independent random variables emanated from the probability density function of $f_X(x)$.

The sample mean, $\overline{\mu_X}$, is obtained as;

$$\overline{\mu_X} = \frac{1}{n} \sum_{i=1}^{n} X_i$$

which is employed to estimate the expected value of the corresponding random process.

The sample variance, $\overline{\sigma_X^2}$, which indicates the average dispersion of the measured values around the sample mean, can be derived as;

© The Author(s), under exclusive license to Springer Nature Switzerland AG 2022
M. Catak et al., *Probability and Random Variables for Electrical Engineering*,
Studies in Systems, Decision and Control 390,
https://doi.org/10.1007/978-3-030-82922-3_9

$$\overline{\sigma_X^2} = \frac{1}{n-1} \sum_{i=1}^{n} (X_i - \overline{\mu_X})^2$$

The coefficient of variation c_X, that measures the variety of the samples around the sample mean, can be formulated as;

$$c_X = \frac{\overline{\sigma_X}}{\mu_X}$$

Since it is the ratio between the sample standard deviation and the sample mean, it is dimensionless.

9.2   Random Sample Matrix

In order to investigate the reproducibility and the repeatability of the random experiments aiming to observe the uncertainty level of such a random process, a random matrix containing random samples is created. It is considered that each column (or each row dependable on the appliers) includes a random experiment of the process.

9.2.1   Correlation Matrices

9.2.1.1   Autocorrelation Matrix

As it was discussed in the previous chapter, the autocorrelation between random samples can be explained as;

$$\mathbf{R_{XX}} = E[\mathbf{XX^T}]$$

where T mean the transpose of the sample vector. It can be shown in matrix form as;

$$\mathbf{R_{XX}} = \begin{bmatrix} E[X_1 X_1] & E[X_1 X_2] & \cdots & ; E[X_1 X_n] \\ E[X_2 X_1] & E[X_2 X_2] & \cdots & E[X_2 X_n] \\ \vdots & \vdots & \ddots & \vdots \\ E[X_n X_1] & E[X_n X_2] & \cdots & E[X_n X_n] \end{bmatrix}$$

9.2.1.2 Cross-correlation Matrix

Similarly the cross-correlation matrix of two random processes can be calculated as;

$$\mathbf{R_{XY}} = E[\mathbf{XY^T}]$$

In another notation as matrix form is available;

$$\mathbf{R_{XY}} = \begin{bmatrix} E[X_1Y_1] & E[X_1Y_2] & \cdots & ; E[X_1Y_n] \\ E[X_2Y_1] & E[X_2Y_2] & \cdots & E[X_2Y_n] \\ \vdots & \vdots & \ddots & \vdots \\ E[X_nY_1] & E[X_nY_2] & \cdots & E[X_nY_n] \end{bmatrix}$$

Properties of Correlation Matrix

(1) $\mathbf{R_{XX}}$ is a symmetric matrix for real-valued random variables, i.e. $\mathbf{R_{XX}} = \mathbf{R_{XX}}^T$
(2) $\mathbf{R_{XY}}$ an asymmetric matrix for real-valued random variables, i.e. $\mathbf{R_{XY}} = \mathbf{R_{YX}}^T$
(3) All eigenvalues of $\mathbf{R_{XX}}$ and $\mathbf{R_{XY}}$ are real, and nonnegative.

9.2.2 Covariance Matrices

9.2.2.1 Autocovariance Matrix

The autocovariance function between random samples can be calculated as;

$$\mathbf{C_{XX}} = E[(\mathbf{X} - \overline{\mu_X})(\mathbf{X} - \overline{\mu_X})^T]$$
$$= \mathbf{R_{XX}} - E[\mathbf{X}]E[\mathbf{X}]^T$$
$$= \mathbf{R_{XX}} - \overline{\mu_X}\ \overline{\mu_X}^T$$

or, in matrix form it can be given as;

$$C_{XX} = \begin{bmatrix} E[(X_1 - E[X_1])(X_1 - E[X_1])] & E[(X_1 - E[X_1])(X_2 - E[X_2])] & \cdots & E[(X_1 - E[X_1])(X_n - E[X_n])] \\ E[(X_2 - E[X_2])(X_1 - E[X_1])] & E[(X_2 - E[X_2])(X_2 - E[X_2])] & \cdots & E[(X_2 - E[X_2])(X_n - E[X_n])] \\ \vdots & \vdots & \ddots & \vdots \\ E[(X_n - E[X_n])(X_1 - E[X_1])] & E[(X_n - E[X_n])(X_2 - E[X_2])] & \cdots & E[(X_n - E[X_n])(X_n - E[X_n])] \end{bmatrix}$$

9.2.2.2   Cross-covariance Matrix

The cross-covariance function between random samples can be derived as;

$$\mathbf{C_{XY}} = E[(\mathbf{X} - \overline{\mu_X})(\mathbf{Y} - \overline{\mu_Y})^T]$$
$$= \mathbf{R_{XY}} - E[\mathbf{X}]E[\mathbf{Y}]^T$$
$$= \mathbf{R_{XX}} - \overline{\mu_X}\ \overline{\mu_Y}^T$$

In matrix form it reads as;

$$\mathbf{C_{XY}} = \begin{bmatrix} E[(X_1 - E[X_1])(Y_1 - E[Y_1])] & E[(X_1 - E[X_1])(Y_2 - E[Y_2])] & \cdots & E[(X_1 - E[X_1])(Y_n - E[Y_n])] \\ E[(X_2 - E[X_2])(Y_1 - E[Y_1])] & E[(X_2 - E[X_2])(Y_2 - E[Y_2])] & \cdots & E[(X_2 - E[X_2])(Y_n - E[Y_n])] \\ \vdots & \vdots & \ddots & \vdots \\ E[(X_n - E[X_n])(Y_1 - E[Y_1])] & E[(X_n - E[X_n])(Y_2 - E[Y_2])] & \cdots & E[(X_n - E[X_n])(Y_n - E[Y_n])] \end{bmatrix}$$

Properties of Correlation Matrix

(1) $\mathbf{C_{XX}}$ is a symmetric matrix for real-valued random variables, i.e.

$$\mathbf{C_{XX}} = \mathbf{C_{XX}}^T$$

(2) $\mathbf{C_{XY}}$ an asymmetric matrix for real-valued random variables, i.e.

$$\mathbf{C_{XY}} = \mathbf{C_{YX}}^T$$

Example 9.1 Consider the random samples given in Table 9.1 where each column is a random sequence. The random samples generated from gaussian distribution with $\mu = 47$, and $\sigma = 3$ using GNU Octave.

(a) Calculate the sample mean, $\overline{\mu_X}$
(b) Calculate the sample standard deviation, $\overline{\sigma_X}$
(c) Find the coefficient of variation, c_X, for each random sequence
(d) Derive the autocorrelation matrix, $\mathbf{R_{XX}}$
(e) Derive the autocovariance matrix, $\mathbf{C_{XX}}$

Solution 9.1 (a) For each random sequence, the sample mean is calculated as;

$$\overline{\mu_X} = \frac{1}{20}\sum_{i=1}^{20} X_i$$

therefore, the sample mean vector, which is a $1 \times n$ matrix;

$$\overline{\mu_X} = \begin{bmatrix} 47.3120 & 47.3040 & 46.8244 & 46.2077 & 45.7533 \end{bmatrix}$$

Table 9.1 Random samples generated from gaussian distribution with $\mu = 47$, and $\sigma = 3$

X_1	X_2	X_3	X_4	X_5
42.491	41.845	45.814	46.233	44.595
50.097	49.661	45.544	51.851	43.982
46.41	46.731	46.178	49.278	47.749
48.877	44.072	46.625	46.12	46.127
43.063	42.223	47.091	48.112	44.263
49.667	45.938	46.421	43.879	46.958
44.781	47.867	48.148	45.739	37.41
46.796	40.434	51.09	44.826	42.125
50.209	46.891	49.673	46.962	45.454
46.522	48.295	47.06	46.115	44.423
46.679	47.947	50.039	44.457	46.95
46.232	45.208	47.125	45.432	47.402
45.172	46.897	41.746	42.771	49.037
50.436	53.024	41.094	45.346	47.348
51.457	46.634	48.287	49.12	42.922
48.124	46.269	46.434	47.659	50.592
45.112	52.501	43.39	48.85	46.484
46.257	46.97	52.498	40.886	46.54
47.449	50.337	44.625	46.127	50.69
50.409	50.863	47.607	44.391	44.013

(b) For each random sequence, the sample standard deviation is obtained as;

$$\overline{\sigma_X} = \sqrt{\frac{1}{n-1} \sum_{i=1}^{n} (X_i - \overline{\mu_X})^2}$$

then, the sample standard deviation vector is;

$$\overline{\sigma_\mathbf{X}} = \begin{bmatrix} 2.5429 & 3.3179 & 2.8306 & 2.4968 & 3.0294 \end{bmatrix}$$

(c) The coefficient of variation is calculated as;

$$c_X = \frac{\overline{\sigma_X}}{\overline{\mu_X}}$$

then it is

$$\mathbf{c_X} = \begin{bmatrix} 0.0537 & 0.0701 & 0.0605 & 0.0540 & 0.0662 \end{bmatrix}$$

Table 9.2 Random samples generated from a Poisson distribution with $\lambda = 5$

X_1	1	6	8	7	1	4	8	9	7	3	2	5	6	10	6	3	6	6	5	7
X_2	8	6	4	0	5	3	6	7	10	10	2	7	9	3	6	6	6	7	5	2

(d) The autocorrelation matrix is obtained as follows;

$$\mathbf{R_{XX}} = \begin{bmatrix} 3.1248 & 0.18303 & -0.22878 & 0.43724 & 0.84829 \\ 0.18303 & 3.0309 & 0.041451 & 0.066918 & 0.7668 \\ -0.22878 & 0.041451 & 2.2750 & 0.33994 & -1.0113 \\ 0.43724 & 0.066918 & 0.33994 & 3.2480 & 0.32915 \\ 0.84829 & 0.7668 & -1.0113 & 0.32915 & 3.2407 \end{bmatrix}$$

(e) The autocovariance matrix is obtained as;

$$\mathbf{C_{XX}} = \begin{bmatrix} 7.0234 & -3.8679 & 1.5705 & -0.78335 & -1.4386 \\ -3.8679 & 10.446 & 2.4055 & -1.0038 & 3.108 \\ 1.5705 & 2.4055 & 10.378 & -0.051767 & 0.73686 \\ -0.78335 & -1.0038 & -0.051767 & 3.8021 & -2.0044 \\ -1.4386 & 3.108 & 0.73686 & -2.0044 & 5.5669 \end{bmatrix}$$

9.2.3 Sample Correlation Coefficient

The correlation coefficient between two random sequences can be calculated as;

$$\overline{\rho_{XX}}(k) = \frac{1}{(n-k)\overline{\sigma_X}^2} \sum_{i=1}^{n-k} (X_i - \overline{\mu_X})(X_{i+k} - \overline{\mu_X})$$

In practical applications, the correlation coefficient can be employed in order to investigate the similarity level of two random sequences at the corresponding time-shift k.

Example 9.2 The random samples given in Table 9.2 were generated from a Poisson distribution with $\lambda = 5$. Assume that they are random experimental data representing two random sequences. Calculate the correlation coefficient as a function of time-shift k.

Solution 9.2 The correlation coefficients are calculated as considering $-19 \le k \le 19$, since the size of the random sequence is 20×1. The results are depicted in Fig. 9.1. According to the results, $k = 1$ gives the maximum correlation between the random sequences.

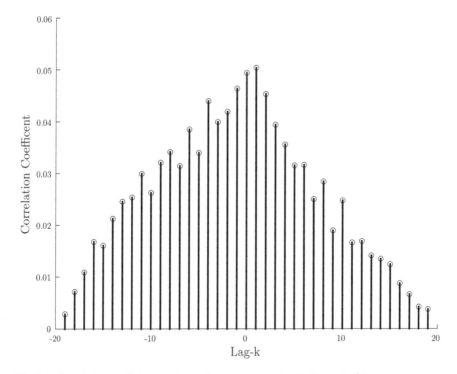

Fig. 9.1 Correlation coefficients of the random sequences given in Example 9.2

9.3 Confidence Interval

Assuming that the random sequence $\mathbf{X} = [X_1, X_2, \ldots, X_n]$ is obtained from a distribution with the mean μ_X, and the standard deviation of σ_X^2. According to the central limit theorem, if n is very large (in theory if $n \to \infty$) the distribution belongs to the sample mean, $\overline{\mu_X}$, and the sample standard deviation, $\overline{\sigma_X}$, can be approximately modelled by means of a normal distribution, such that

$$\sqrt{n}\frac{\overline{\mu_X} - \mu_X}{\overline{\sigma_X}} \approx N(0, 1)$$

where $N(0, 1)$ is the standard normal distribution.

Considering a parameter α, where $0 < \alpha < 1$, then x_α states that;

$$P(X > x_\alpha) = \alpha$$

Using the symmetry property of a normal distribution, it can be written that

$$P\left(-x_{\frac{\alpha}{2}} < X < x_{\frac{\alpha}{2}}\right) = 1 - \alpha$$

or, alternatively

$$P\left(\overline{\mu_X} - x_{\frac{\alpha}{2}}\frac{\overline{\sigma_X}}{\sqrt{n}} < \mu_X < \overline{\mu_X} + x_{\frac{\alpha}{2}}\frac{\overline{\sigma_X}}{\sqrt{n}}\right) = 1 - \alpha$$

This means that the expected value of μ_X can be represented by the sample mean $\overline{\mu_X}$ with the percentage of $(1 - \alpha) \times 100$, which is called confidence interval of the random experiment.

Example 9.3 Let X be a random variable representing the current passing through a resistor in a linear circuit. Suppose that, the current has been measured 64 times, then the sample mean, and the sample standard deviation are calculated as $\overline{\mu_X} = 2.05$ mA, $\overline{\sigma_X} = 0.3\ mA$. Find the 95% confidence interval for the current.

Solution 9.3 Since α is given as 0.05, then using standard normal distribution Figure given in Sect. 9.2, $x_{\frac{\alpha}{2}} = 1.96$.
By employing,

$$P\left(2.05 - 1.96\frac{0.3}{\sqrt{64}} < \mu_X < 2.05 + 1.96\frac{0.3}{\sqrt{64}}\right) = 0.95$$

The corresponding interval is obtained as $[1.9765, 2.1235]$, which means, μ_X will be lie in the interval of $[1.9765, 2.1235]$ with 95%.

9.4   Linear Transformation of Gaussian Random Variables

Suppose that $\mathbf{X} = [X_1, X_2, \ldots, X_n]$ are random variables having known statistical parameters,such as $\overline{\mu_X}$, $\overline{\sigma_X}$, R_{XX}, etc. Let define a new random vector$\mathbf{Y}$, which is defined linear combination of the random variables of $[X_1, X_2, \ldots, X_n]$;

$$\mathbf{Y} = \mathbf{AX}$$

Foe the sake of the simplicity, we discuss the 2×2 case;

$$Y_1 = aX_1 + bX_2$$
$$Y_2 = cX_1 + dX_2$$

or in matrix form;

$$\mathbf{Y} = \begin{bmatrix} a & b \\ c & d \end{bmatrix} \mathbf{X}$$

The expected value of $\mathbf{Y}$ can be calculated as;

$$E[Y_1] = E[aX_1 + bX_2] = aE[X_1] + bE[X_2]$$
$$E[Y_2] = E[cX_1 + dX_2] = cE[X_1] + dE[X_2]$$

Therefore it can be written as;

$$\mu_Y = A\mu_X$$

The variance and the covariance calculations can be derived by;

$$Var[Y_1] = Var[aX_1 + bX_2] = a^2Var[X_1] + b^2Var[X_2] + 2abCov[X_1, X_2]$$
$$Var[Y_2] = Var[cX_1 + dX_2] = c^2Var[X_1] + d^2Var[X_2] + 2cdCov[X_1, X_2]$$
$$Cov[Y_1, Y_2] = Cov[(aX_1 + bX_2), (cX_1 + dX_2)]$$
$$= acVar[X_1] + bdVar[X_2] + adCov[X_1, X_2] + bdCov[X_1, X_2]$$

Finally, the covariance matrix of **Y**, $\mathbf{C_{YY}}$ can be expressed as;

$$\mathbf{C_{YY}} = \mathbf{AC_{XX}A}^T$$

where T stands for the transpose of the corresponding matrix.

Example 9.4 Considering X_1 and X_2 are two random variables having $\mu_{X_1} = 2$, $\mu_{X_2} = 3$ and the following covariance matrix;

$$\mathbf{C_{XX}} = \begin{bmatrix} 5 & 2 \\ 2 & 6 \end{bmatrix}$$

Let Y_1 and Y_2 be expressed as linear transformation of X_1 and X_2 such that;

$$Y_1 = 2X_1 + 3X_2$$
$$Y_2 = -5X_1 + X_2$$

(a) Derive μ_X
(b) Calculate $\mathbf{C_{YY}}$
(c) Find $Var[Y_1]$, and $Var[Y_2]$

Solution 9.4

(a) The linear transformation matrix is given as;

$$\mathbf{A} = \begin{bmatrix} 2 & 3 \\ -5 & 1 \end{bmatrix}$$

$$\mu_Y = A\mu_X$$

$$= \begin{bmatrix} 2 & 3 \\ -5 & 1 \end{bmatrix} \begin{bmatrix} 2 \\ 3 \end{bmatrix}$$

$$= \begin{bmatrix} 13 \\ -7 \end{bmatrix}$$

Then, $\mu_{Y_1} = 13$, and $\mu_{Y_2} = -7$

(b)

$$C_{YY} = AC_{XX}A^T$$

$$= \begin{bmatrix} 2 & 3 \\ -5 & 1 \end{bmatrix} \begin{bmatrix} 5 & 2 \\ 2 & 6 \end{bmatrix} \begin{bmatrix} 2 & -5 \\ 3 & 1 \end{bmatrix}$$

$$= \begin{bmatrix} 98 & -58 \\ -58 & 111 \end{bmatrix}$$

(c) Using the results obtained in part (b), $Var[Y_1] = 98$ and $Var[Y_2] = 111$.

Example 9.5 Considering the circuit shown in Fig. 9.2 where $R_1 = 6\ k\Omega$, $R_2 = 3$ $k\Omega$, $R_3 = 2\ k\Omega$, $R_{f1} = 12\ k\Omega$, $R_4 = 5\ k\Omega$, and $R_{f2} = 10\ k\Omega$.

(a) Derive the transformation matrix A such that $Y = AX$
   If the inputs have

$$\mu_X = \begin{bmatrix} 2 \\ 1 \\ 3 \end{bmatrix}$$

$$C_{XX} = \begin{bmatrix} 4 & 2 & 1 \\ 2 & 6 & 3 \\ 1 & 3 & 8 \end{bmatrix}$$

(b) Calculate C_{YY}

Solution 9.5

(a) The first part of the circuit is a summing amplifier, then its output can be written as;

$$Y_1 = \left(\frac{R_{f1}}{R_1} X_1 + \frac{R_{f1}}{R_2} X_2 + \frac{R_{f1}}{R_3} X_3 \right)$$

$$= -2X_1 - 4X_2 - 6X_3$$

The second part of the circuit is a inverting amplifier, its output can be expressed as;

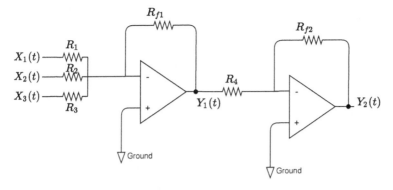

Fig. 9.2 A cascaded op-amp circuit

$$Y_2 = \left(\frac{R_{f2}}{R_4}Y_1\right)$$
$$= -2Y_1$$
$$= 4X_1 + 8X_2 + 12X_3$$

Therefore the transformation matrix is;

$$\mathbf{A} = \begin{bmatrix} -2 & -4 & -6 \\ 4 & 8 & 12 \end{bmatrix}$$

(b)

$$\mathbf{C_{YY}} = \mathbf{AC_{XX}A}^T$$

$$= \begin{bmatrix} -2 & -4 & -6 \\ 4 & 8 & 12 \end{bmatrix} \begin{bmatrix} 4 & 2 & 1 \\ 2 & 6 & 3 \\ 1 & 3 & 8 \end{bmatrix} \begin{bmatrix} -2 & 4 \\ -4 & 8 \\ -6 & 12 \end{bmatrix}$$

$$= \begin{bmatrix} 600 & -1200 \\ -1200 & 2400 \end{bmatrix}.$$

9.5 Problems

9.1 Consider the random samples given in Table 9.3 where each column is a random sequence. The random samples generated from gaussian distribution with $\mu = 68$, and $\sigma = 5$ using GNU Octave.

(a) Calculate the sample mean, $\overline{\mu_X}$
(b) Calculate the sample standard deviation, $\overline{\sigma_X}$
(c) Find the coefficient of variation, c_X, for each random sequence

Table 9.3 Random samples generated from gaussian distribution with $\mu = 68$, and $\sigma = 5$

X_1	X_2	X_3	X_4	X_5	X_6	X_7	X_8	X_9	X_{10}
65.03	75.334	67.395	63.67	78.939	62.763	65.983	64.473	62.574	71.011
62.715	69.12	71.378	62.565	75.67	67.558	66.185	64.432	70.459	70.542
70.013	64.774	64.507	71.353	54.988	78.291	70.217	58.916	62.367	67.898
63.747	65.316	68.531	63.808	62.861	68.105	68.687	64.535	62.378	66.549
68.539	65.159	63.884	72.731	62.937	67.377	72.403	67.017	65.17	68.41
69.725	64.209	67.703	57.284	69.002	67.871	67.606	67.718	67.47	64.953
67.168	76.625	59.836	60.932	66.013	62.019	68.439	58.888	65.015	70.629
58.466	60.122	70.728	72.079	64.552	66.917	61.905	75.602	64.752	63.387
71.718	61.466	65.455	80.182	65.174	75.137	76.38	64.197	78.756	71.876
71.669	66.834	66.833	62.411	63.592	64.222	68.16	61.715	68.409	69.753
64.963	59.418	71.122	57.376	67.701	68.376	71.165	68.076	67.878	59.801
71.763	69.856	61.449	64.459	69.231	68.854	69.401	61.204	72.512	64.836
66.086	60.788	59.033	73.528	66.45	67.07	65.401	67.232	67.928	66.664
67.336	60.527	70.728	69.438	67.514	69.76	67.17	66.443	75.041	76.663
69.746	71.076	66.378	70.078	76.586	64.193	64.361	67.921	63.989	71.433
74.832	68.611	68.184	62.209	67.996	71.953	67.772	67.529	63.707	72.388
67.859	68.428	68.091	66.313	64.015	63.71	67.373	74.866	65.561	65.863
70.369	62.326	65.656	58.382	75.332	68.011	59.547	58.454	70.056	70.659
63.954	74.861	69.014	70.37	66.507	73.167	79.357	64.077	67.074	68.248
68.088	69.638	69.114	70.38	69.456	67.454	72.906	70.966	67.356	67.193
64.782	63.7	68.488	74.238	52.792	72.299	67.625	60.438	65.12	58.326
65.086	69.088	63.961	74.055	56.364	73.741	65.483	63.01	69.917	69.272
65.27	69.282	73.917	73.898	63.335	70.713	60.399	64.628	75.227	54.723
74.831	68.665	62.867	62.809	75.678	63.767	61.099	69.258	69.624	75.931
69.935	69.3	51.363	68.626	61.861	66.38	67.826	59.846	72.389	68.44

Table 9.4 Random samples generated from a Poisson distribution with $\lambda = 10$

X_1	8	11	9	14	7	8	7	8	9	7	9	12	6	10	8	14	6	9	14	18	12	13	8	8	9	7	7	7	10	9
X_2	14	6	6	13	7	9	10	15	8	5	7	10	13	6	5	7	9	8	3	10	9	10	12	15	9	8	7	11	12	14

(d) Derive the autocorrelation matrix, $\mathbf{R_{XX}}$

(e) Derive the autocovariance matrix, $\mathbf{C_{XX}}$.

9.2 The random samples given in Table 9.4 were generated from a Poisson distribution with $\lambda = 10$. Assume that they are random experimental data representing two random sequences.

(a) Calculate the correlation coefficient as a function of time-shift k.
(b) Find the value of k gives the maximum correlation between the random sequences.

9.3 Let X be a random variable representing the potential difference between two ends of a resistor in a linear circuit. Suppose that, the potential difference has been measured 100 times, then the sample mean, and the sample standard deviation are calculated as $\overline{\mu_X} = 5.02\ mA$, $\overline{\sigma_X} = 0.1\ mA$. Find the 95% confidence interval for the potential difference.

9.4 Solve Problem 9.3 considering the 97.5% confidence interval for the potential difference.

9.5 Considering X_1 and X_2 are two random variables having

$$\mu_X = \begin{bmatrix} 4 \\ 0 \\ 3 \end{bmatrix}$$

$$C_{XX} = \begin{bmatrix} 4 & 3 & 1 \\ 3 & 9 & 2 \\ 1 & 2 & 16 \end{bmatrix}$$

If new random sequence matrix **Y** is obtained by means of the following linear transformation;

$$A = \begin{bmatrix} 1 & -1 & 1 \\ 1 & 1 & -1 \\ -1 & 1 & 1 \end{bmatrix}$$

(a) Derive μ_Y
(b) Calculate C_{YY}
(c) Find $Var[X_1]$, $Var[X_2]$, and $Var[X_3]$
(c) Find $Var[Y_1]$, $Var[Y_2]$, and $Var[Y_3]$.

9.6 Consider a linear system having the input and the corresponding measured output data depicted in Table 9.5

(a) Derive $\overline{\mu_Y}$
(b) Calculate $\overline{\sigma_Y}^2$
(c) Find the correlation between X and Y; R_{XY}.

9.7 Solve the problem given in Example 9.5, where $R_1 = 1\ k\Omega$, $R_2 = 2\ k\Omega$, $R_3 = 4\ k\Omega$, $R_{f1} = 16\ k\Omega$, $R_4 = 6\ k\Omega$, and $R_{f2} = 3\ k\Omega$.

Table 9.5 A linear system having the input and the corresponding measured output data

X, input	Y, output
0.1	2.3987
0.2	2.8199
0.3	3.1918
0.4	3.6208
0.5	4.0403
0.6	4.3689
0.7	4.7583
0.8	5.1737
0.9	5.6293
1	5.9334
1.1	6.4156
1.2	6.87
1.3	7.1074
1.4	7.6822
1.5	8.1695

Appendix A
Fourier Transform

A.1 Continuous Time Fourier Transform, CTFT

The Fourier transform of a function $x(t)$ can be derived as;

$$\mathcal{F}\{x(t)\} = X(jw)$$

$$= \int_{-\infty}^{\infty} x(t)e^{-jwt}\,dt$$

The function of $X(jw)$ is the spectrum of $x(t)$, which is the corresponding function in frequency domain. For some cases, if $X(jw)$ is known, then $x(t)$ can be derived using inverse Fourier transform;

$$x(t) = \mathcal{F}^{-1}\{X(jw)\}$$

$$= \frac{1}{2\pi} \int_{-\infty}^{\infty} X(jw)e^{jwt}\,dw$$

A.1.1 Properties of CTFT

A.1.1.1 Linearity

For any constant value of α;

(i) $\mathcal{F}\{\alpha x(t)\} = \alpha X(jw)$
(ii) $\mathcal{F}\{x_1(t) + x_2(t)\} = \mathcal{F}\{x_1(t)\} + \mathcal{F}\{x_2(t)\}$

© The Editor(s) (if applicable) and The Author(s), under exclusive license
to Springer Nature Switzerland AG 2022
M. Catak et al., *Probability and Random Variables for Electrical Engineering*,
Studies in Systems, Decision and Control 390,
https://doi.org/10.1007/978-3-030-82922-3

In general it can be written as;

$$X(jw) = \mathcal{F}\left\{\sum_{i=1}^{\infty}\alpha_i x_i(t)\right\} = \sum_{i=1}^{\infty}\alpha_i \mathcal{F}\{x_i(t)\} = \sum_{i=1}^{\infty}\alpha_i X_i(jw)$$

A.1.1.2　Shifting in Time

If it is known that

$$x(t) \leftrightarrow X(jw)$$

Considering a time shift t_0, we have

$$x(t - t_0) \leftrightarrow X(jw)e^{-jwt_0}$$

A.1.1.3　Shifting in Frequency

If it is known that

$$x(t) \leftrightarrow X(jw)$$

Considering a frequency shift of w_0, one has

$$x(t)e^{jw_0 t} \leftrightarrow X(j(w - w_0))$$

A.1.1.4　Scaling

Assuming that α is a real constant, then

$$x(\alpha t) \leftrightarrow \frac{1}{|\alpha|}X\left(\frac{jw}{\alpha}\right)$$

A.1.1.5　Differentiation in Time Domain

Assuming that $x(t)$ is a continuous and differentiable function in the interested time interval,

$$\frac{d^n x(t)}{dt^n} \leftrightarrow (jw)^n X(jw)$$

A.1.1.6 Differentiation in Frequency Domain

Assuming that $X(jw)$ is a continuous and differentiable function in the interested frequency interval,

$$\frac{\mathrm{d}^n X(jw)}{\mathrm{d}w^n} \leftrightarrow (-jt)^n x(t)$$

A.1.1.7 Convolution in Time Domain

Assuming that

$$x_1(t) \leftrightarrow X_1(jw)$$
$$x_2(t) \leftrightarrow X_2(jw)$$

then,

$$\int_{-\infty}^{\infty} x_1(\tau)x_2(t-\tau)\mathrm{d}\tau \leftrightarrow X_1(jw)X_2(jw)$$

A.1.1.8 Multiplication in Time Domain or Convolution in Frequency Domain

Assuming that

$$x_1(t) \leftrightarrow X_1(jw)$$
$$x_2(t) \leftrightarrow X_2(jw)$$

then,

$$x_1(t)x_2(t) \leftrightarrow \int_{-\infty}^{\infty} X_1(\zeta)X_2(w-\zeta)\mathrm{d}\zeta$$

A.2 Some Commonly Used CTFT Pairs

Some commonly used CTFT pairs are shown in Table A.1.

A.3 Discrete Time Fourier Transform, DTFT

Suppose that $x[n]$ is a discrete time function, then Discrete Time Fourier Transform (DTFT) of $x[n]$ can be derived as;

Table A.1 Some commonly used CTFT pairs

$x(t) = \frac{1}{2\pi} \int_{-\infty}^{\infty} X(j\omega) e^{j\omega t} d\omega$	$X(j\omega) = \int_{-\infty}^{\infty} x(t) e^{-j\omega t} dt$
$x(t - t_0)$	$X(j\omega) e^{-j\omega t_0}$
$x(t) e^{j\omega_0 t}$	$X(j(\omega - \omega_0))$
$x(\alpha t)$	$\frac{1}{\|\alpha\|} X\left(\frac{j\omega}{\alpha}\right)$
$X(t)$	$2\pi x(-j\omega)$
$\frac{d^n x(t)}{dt^n}$	$(j\omega)^n X(j\omega)$
$(-jt)^n x(t)$	$\frac{d^n X(j\omega)}{d\omega^n}$
$\int_{-\infty}^{t} x(\tau) d\tau$	$\frac{X(j\omega)}{j\omega} + \pi F(0)\delta(\omega)$
$\delta(t)$	1
$e^{j\omega_0 t}$	$2\pi\delta(\omega - \omega_0)$
$\text{sgn}(t)$	$\frac{2}{j\omega}$
$j\frac{1}{\pi t}$	$\text{sgn}(\omega)$
$u(t)$	$\pi\delta(\omega) + \frac{1}{j\omega}$
$\sum_{n=-\infty}^{\infty} F_n e^{jn\omega_0 t}$	$2\pi \sum_{n=-\infty}^{\infty} F_n \delta(\omega - n\omega_0)$
$A\cos\left(\frac{\pi t}{2\tau}\right) \text{rect}\left(\frac{t}{2\tau}\right)$	$\frac{A\pi}{\tau} \frac{\cos(\omega\tau)}{(\pi/2\tau^2 - \omega^2}$
$\cos(\omega_0 t)$	$\pi[\delta(\omega - \omega_0) + \delta(\omega + \omega_0)]$
$\sin(\omega_0 t)$	$\frac{\pi}{j}[\delta(\omega - \omega_0) - \delta(\omega + \omega_0)]$
$u(t)\cos(\omega_0 t)$	$\frac{\pi}{2}[\delta(\omega - \omega_0) + \delta(\omega + \omega_0)] + \frac{j\omega}{\omega_0^2 - \omega^2}$
$u(t)\sin(\omega_0 t)$	$\frac{\pi}{2j}[\delta(\omega - \omega_0) - \delta(\omega + \omega_0)] + \frac{\omega^2}{\omega_0^2 - \omega^2}$
$u(t)e^{-at}\cos(\omega_0 t)$	$\frac{(\alpha + j\omega)}{\omega_0^2 + (\alpha + j\omega)^2}$
$u(t)e^{-\alpha t}\sin(\omega_0 t)$	$\frac{\omega_0}{\omega_0^2 + (\alpha + j\omega)^2}$
$e^{-\alpha\|t\|}$	$\frac{2\alpha}{\alpha^2 + \omega^2}$
$e^{-t^2/(2\sigma^2)}$	$\sigma\sqrt{2\pi} e^{-\sigma^2\omega^2/2}$
$u(t)e^{-at}$	$\frac{1}{\alpha + j\omega}$
$u(t)te^{-\alpha t}$	$\frac{1}{(\alpha + j\omega)^2}$

$$X(e^{jw}) = \sum_{n=-\infty}^{\infty} x[n] e^{-jwn}$$

and it is notated as

$$X(e^{jw}) = \mathcal{F}^{-1}\{x[n]\}$$

If $X(e^{jw})$ is known, then $x[n]$ can be derived by means of Inverse DTFT, such that

$$x[n] = \frac{1}{2\pi} \int_{-\pi}^{\pi} X(e^{jw}) e^{jwn} dw$$

In general, $x[n]$ and $X(e^{jw})$ are called DTFT pair, then the relationship is notated as;

$$x[n] \leftrightarrow X(e^{jw}).$$

A.3.1 Properties of DTFT

A.3.1.1 Linearity

For any constant value of α;

(i) $\mathcal{F}\{\alpha x[n]\} = \alpha X(e^{jw})$

(ii) $\mathcal{F}\{x_1[n] + x_2[n]\} = \mathcal{F}\{x_1[n]\} + \mathcal{F}\{x_2[n]\}$

In general it can be written as;

$$X(e^{jw}) = \mathcal{F}\left\{\sum_{i=1}^{\infty} \alpha_i x_i[n]\right\} = \sum_{i=1}^{\infty} \alpha_i \mathcal{F}\{x_i[n]\} = \sum_{i=1}^{\infty} \alpha_i X_i(e^{jw})$$

A.3.1.2 Shifting in Time

If it is known that

$$x[n] \leftrightarrow X(e^{jw})$$

Considering a time shift n_0, this yields

$$x[n - n_0] \leftrightarrow X(e^{jw})e^{-jwn_0}$$

A.3.1.3 Shifting in Frequency

If it is given that

$$x[n] \leftrightarrow X(e^{jw})$$

Considering a frequency shift of w_0, one has

$$x[n]e^{jw_0 n} \leftrightarrow X(e^{j(w-w_0)})$$

A.3.1.4   Differentiation in Frequency Domain

Assuming that $X(e^{jw})$ is a continuous and differentiable function,

$$j\frac{dX(e^{jw})}{dw} \leftrightarrow nx[n]$$

A.3.1.5   Convolution Sum in Time Domain

Assuming that

$$x_1[n] \leftrightarrow X_1(e^{jw})$$

$$x_2[n] \leftrightarrow X_2(e^{jw})$$

and $y[n]$ is defined as the convolution sum of $x_1[n]$ and $x_2[n]$;

$$y[n] = x_1[n] * x_2[n]$$

$$= \sum_{k=-\infty}^{\infty} x_1[k]x_2[n-k]$$

then it can be written as;

$$Y(e^{jw}) = X_1(e^{jw})X_2(e^{jw})$$

A.3.1.6   Multiplication in Time Domain or Convolution in Frequency Domain

Assuming that

$$x_1[n] \leftrightarrow X_1(e^{jw})$$

$$x_2[n] \leftrightarrow X_2(e^{jw})$$

then,

$$x_1[n]x_2[n] \leftrightarrow \frac{1}{2\pi} \int_{-\pi}^{\pi} X_1(e^{j\zeta})X_2(e^{j(w-\zeta)})d\zeta.$$

A.4   Some Commonly Used DTFT Pairs

Some commonly used DTFT pairs are shown in Table A.2.

Table A.2 Some commonly used DTFT pairs

$x[n] = \delta[n]$	$X\left(e^{j\omega}\right) = 1$
$x[n] = \delta[n - n_0]$	$X\left(e^{j\omega}\right) = e^{-jwn_0}$
$x[n] = \alpha^n u[n], \quad |\alpha| < 1$	$X\left(e^{j\omega}\right) = \frac{1}{1-\alpha e^{-j\omega}}$
$x[n] = u[n]$	$X\left(e^{j\omega}\right) = \frac{1}{1-e^{-j\omega}} + \pi \sum_{k=-\infty}^{\infty} \delta(\omega + 2\pi k)$
$x[n] = \begin{cases} 1, 0 \leq n \leq M \\ 0, \text{ otherwise} \end{cases}$	$X\left(e^{j\omega}\right) = \frac{\sin\left[\omega\left(\frac{M+1}{2}\right)\right]}{\sin\left(\frac{\omega}{2}\right)} e^{-\frac{jwM}{2}}$
$x[n] = \frac{\sin(w_c n)}{\pi n}, 0 < w_c \leq \pi$	$X\left(e^{j\omega}\right) = \begin{cases} 1, |\omega| < w_c \\ 0, w_c < |\omega| \leq \pi \end{cases}$
$x[n] = (n+1)\alpha^n u[n]$	$X\left(e^{j\omega}\right) = \frac{1}{\left(1-\alpha e^{-j\omega}\right)^2}$

Appendix B
Probability Distribution Functions, Summary

B.1 Continuous Time Distributions

B.1.1 Uniform Distribution

$$f_X(x) = \frac{1}{b-a}$$

$$E[X] = \frac{a+b}{2}$$

$$Var[X] = \frac{1}{12}(b-a)^2$$

B.1.2 Normal (Gaussian) Distribution

The mean (μ_X), and the standard deviation (σ_X), characterize a normal distribution. A univariate normal probability density function is expressed as;

$$f_X(x) = \frac{1}{\sqrt{2\pi\sigma_X^2}} e^{\frac{-1}{2}\left(\frac{x-\mu_X}{\sigma_X}\right)^2}$$

© The Editor(s) (if applicable) and The Author(s), under exclusive license to Springer Nature Switzerland AG 2022
M. Catak et al., *Probability and Random Variables for Electrical Engineering*, Studies in Systems, Decision and Control 390, https://doi.org/10.1007/978-3-030-82922-3

B.1.3   Exponential Distribution

$$f_X(x) = \begin{cases} \lambda e^{\lambda x} & \text{for } x \geq 0 \\ 0 & \text{elsewhere} \end{cases}$$

where λ is a proper positive constant.

$$E[X] = \frac{1}{\lambda}$$

$$\text{Var}[X] = \frac{1}{\lambda^2}$$

B.1.4   Gamma Distribution

$$f_X(x) = \begin{cases} \frac{\lambda^{\alpha} x^{\alpha-1} e^{-\lambda x}}{\Gamma(\alpha)} & x > 0 \\ 0 & \text{elsewhere} \end{cases}$$

where $\lambda > 0$, and $\alpha > 0$. $\Gamma(\alpha)$ is called Gamma function that equals;

$$\Gamma(\alpha) = \int_0^{\infty} x^{\alpha-1} e^{-x} dx$$

The expected value of a gamma distribution is;

$$E[X] = \frac{\alpha}{\lambda}$$

and the variance of a gamma distribution is;

$$\text{Var}[X] = \sigma_X^2 = \frac{\alpha}{\lambda^2}$$

B.2 Discrete Time Distributions

B.2.1 Uniform Distribution

$$f_X(x) = \sum_{i=1}^{n} \frac{1}{n} \delta(x - x_i)$$

$$E[X] = \sum_{i=a}^{b} i \frac{1}{b-a+1}$$

$$= \frac{1}{b-a+1} \sum_{i=a}^{b} i$$

$$= \frac{1}{b-a+1} \left(\sum_{i=1}^{b} i - \sum_{i=1}^{a} i \right)$$

$$= \frac{b(b+1) - a(a-1)}{2(b-a+1)}$$

$$= \frac{a+b}{2}$$

$$\text{Var}[X] = \frac{(b-a+1)^2 - 1}{12}$$

B.2.2 Bernoulli Distribution

$$f_X(x) = \begin{cases} p, & x = 0 \\ q, & x = 1 \end{cases}$$

where $p \geq 0, q \geq 0$, and $p + q = 1$ It can be written as;

$$f_X(x) = p\delta(x) + q\delta(x - 1)$$

$$E[X] = p$$

$$\text{Var}[X] = pq = p(1 - p)$$

B.2.3 Binomial Distribution

$$f_X(x) = \sum_{x_i} \binom{n}{x} p^x (1 - p)^{n-x} \delta\,(x - x_i)$$

$$E[X] = np$$
$$\text{Var}[X] = npq = np(1 - p)$$

B.2.4 Geometric Distribution

$$f_X(x) = \sum_{x_i} (1 - p)^{x-1} p \delta\,(x - x_i) \quad , \quad x_i = 1, 2, \ldots$$

$$E[X] = \frac{1}{p}$$

$$\text{Var}[X] = \frac{1 - p}{p^2}$$

B.2.5 Poisson Distribution

$$f_X(x) = e^{-\lambda} \sum_{k=0}^{\infty} \frac{\lambda^k}{k!} \delta(x - k)$$

$$F_X(x) = e^{-\lambda} \sum_{k=0}^{\infty} \frac{\lambda^k}{k!} u(x - k)$$

$$E[X] = \lambda$$
$$\text{Var}[X] = \lambda$$

Bibliography

1. Kay, S.M.: Fundamentals of Statistical Signal Processing. Prentice Hall PTR (1993)
2. Leon-Garcia, A.: Probability and Random Processes for Electrical Engineering. Pearson Education India (1994)
3. Oppenheim, A.V., Buck, J.R., Schafer, R.W.: Discrete-Time Signal Processing, vol. 2. Prentice Hall, Upper Saddle River, NJ (2001)
4. Papoulis, A., Saunders, H.: Probability, Random Variables and Stochastic Processes (1989)
5. Kun Il Park and Park: Fundamentals of Probability and Stochastic Processes with Applications to Communications. Springer (2018)
6. Peebles, P.Z.: Probability, Random Variables, and Random Signal Principles. McGraw Hill (1987)
7. Ross, S.M.: Introduction to Probability and Statistics for Engineers and Scientists. Elsevier (2004)
8. Stark, H., Woods, J.W.: Probability, Random Processes, and Estimation Theory for Engineers. Prentice-Hall Inc (1986)

Index

A

Autocorrelation, 101–104, 107, 109, 112–114, 119, 120, 122–124, 131–139, 142, 144, 146, 152

Autocovariance, 106, 109, 143, 144, 146, 152

B

Bernoulli, 46–49, 165

Bernoulli distribution, 46–48

Binomial, 47, 50, 166

Binomial distribution, 47

C

Central limit theorem, 72, 147

Characteristic function, 81, 82

Complement, 4, 5, 9–11

Conditional probability, 9, 10, 48, 61, 62

Convolution, 69, 70, 128–130, 132, 133, 136, 160

Convolution integral, 70

Convolution sum, 69, 136, 160

Correlation coefficient, 90, 93, 107, 146, 153

Cross-correlation, 104, 105, 132, 133, 143

D

Dirac delta, 41, 42, 46

Discrete random variable, 66, 69

DTFT, 122, 136, 157–161

E

Empty set, 4

Event, 7–11

Expected value, 24, 26, 31, 34, 36, 44, 45, 47, 48, 50, 52, 56, 77–80, 83, 87, 92–94, 101, 102, 104, 107, 109, 114, 118, 119, 130–132, 134, 137, 141, 148, 164

Exponential distribution, 31, 33, 34

F

Fourier transform, 81, 117–120, 129, 133, 134, 155

Frequency shift, 156, 159

G

Gamma, 34, 35, 164

Gamma distribution, 34

Geometric, 48, 166

I

Impulse function, 41, 125, 127

Intersection operation, 4

J

Joint probability distribution, 59–61, 73

M

Maclaurin series, 79

Marginal probability density, 61, 62, 73, 74, 84, 90, 93

Marginal probability distribution, 61, 73

Mean, 24, 26, 29–31, 33, 37

Printed in the United States
by Baker & Taylor Publisher Services